VILLIERS SINGLES
& TWINS

NITON PUBLISHING

VILLIERS SINGLES & TWINS

The postwar British two-stroke lightweight motorcycle

Roy Bacon

Published by Niton Publishing
P.O. Box 3, Ventnor, Isle of Wight, PO38 2AS

© Copyright Roy Bacon 1992

First published in 1983 in Great Britain by Osprey Publishing Limited,
12-14 Long Acre, London WC2E 9LP
Member company of the George Philip Group

This book is copyrighted under the Berne Convention. All rights reserved. Apart from any fair dealing for the purpose of private study, research, criticism or review, as permitted under the Copyright Act, 1956, no part of this publication may be reproduced, stored in a retrieval system, or transmitted in any form or by any means, electronic, electrical, chemical, mechanical, optical, photocopying, recording, or otherwise, without prior written permission. All enquiries should be addressed to the Publisher.

A CIP catalogue record for this book is available from the British Library

ISBN 1 85579 020 3

Original edition :-
Editor Tim Parker
Design Gwyn Lewis
Reprinted by Manor Design & Printing, Isle of Wight

Contents

Acknowledgements 6
Introduction 7

Part One Engines
Villiers 8 / Alpha 20 / AMC 22 / Anzani 23 / Excelsior engines 24 / JAP 27 / RCA 27

Part Two Machines
Aberdale 28 / ABJ 29 / AJS 30 / AJW 31 / Ambassador 31 / BAC 38 / Bond 38 / Bown 40 / Butler 41 / Cheetah 42 / Commander 42 / Corgi 44 / Cotton 47 / Cyc-Auto 51 / DMW 53 / Dot 58 / Elstar 62 / Excelsior 62 / Firefly 69 / FLM 69 / Francis-Barnett 70 / Greeves 81 / HJH 97 / James 98 / Mercury 107 / New Hudson 107 / Norman 109 / OEC 113 / Panther 115 / Radco 117 / Rainbow 117 / Raynal 119 / Sapphire 119 / Scorpion 120 / Sprite 122 / Sun 125 / Tandon 128

Part Three Appendix
1 Villiers engine numbers   131
2 Engine specifications (Villiers, AJS, AMC, Anzani, Excelsior engines, JAP)   138
3 Machine specifications   147
4 Engine and frame numbers   174
5 Villiers carburettor settings   184
6 Amal carburettor settings   185

Acknowledgements

My memories of Villiers engines and the many machines they powered go back to my motorcycling infancy for my local shop was Meetons Motorcycle Mecca at Shannon Corner on the A3. There I listened to Miss Child selling spares to riders who barely knew the make of their machine and was served by Dick Ward who patiently would sort out my problems.

Now the Mecca is buried under a massive interchange, Miss Child is president of the British Two Stroke Club and Dick runs Meeton & Ward down the road at Ewell, still dispensing Villiers spares and knowledge.

This book differs from others in the Osprey Collector's Library series for its theme is the basic transport element as represented by over 30 marques using seven different engines. For information on all these I am indebted to many people and especially to Fluff Brown of AJS, Mike and Frank Cutler of Alpha, Peter Frazer of Murray Evans, Peter Howdell of *Motor Cycle News*, Chris Goodfellow who supplied late Greeves material and of course, Dick Ward and his staff who have always helped me with Villiers queries.

For pictures I again turned to the magazines and am again indebted to Bob Berry and Peter Law of *Motor Cycle News*, Mick Woollett of *Motor Cycle Weekly*, Ian Beacham and Jim Lindsay of *Mechanics*, Bob Currie of *Classic Motor Cycle* and Jeff Clew who lent rare material from his own collection.

One picture came from the Imperial War Museum but most are stock shots taken for catalogues and the firms involved in their production were Bromhead, Industrial Photographic, S. F. James, H. E. Jones, K. G. Jones, Lewis and Randall, Morland Braithwaite, Donald Page, Rimmer, Terric Studios and Arthur Winter.

Other pictures from the *MCN* files were the work of Cecil Bailey and Malcolm Carling. If I have used any others taken by a freelance but unmarked, I can only apologise.

Finally I must once more thank Tim Parker the editor for all his assistance.

Roy Bacon
Niton, Isle of Wight
December 1982

This is another of my *Osprey* titles now published under my *Niton* imprint. The text is essentially as before, with minor corrections, but the tables of engine and other numbers have been expanded to include data that has come to hand since the book was first written.

Niton, January 1992

Introduction

The term 'postwar British two-stroke lightweight motorcycle' embraces what many would consider to be a multitude of sins and at first glance is just about motorcycles with Villiers engines. In fact the spectrum is a little wider for although the bulk of machines of that period were powered by the Wolverhampton make, there were several other engines in popular use.

Excelsior built their own units as well as using Villiers, while the AMC group members—Francis-Barnett and James—introduced considerable confusion by using group AMC engines shortly after DMW used French overhead camshaft four-stroke AMC engines of no relationship at all.

A number of firms used the Anzani twin, while a few fitted the JAP unit. In later years Alpha Bearings came onto the scene and, with several firms offering barrel conversions, moved the power output of the Villiers unit along.

This book set out to cover the Villiers-powered motorcycle and its scope was widened to include all British-powered 'look alikes'. This brought in all the units mentioned above but has kept out certain makes which do not fit the pattern. This, and space considerations, have also kept out all mopeds, as opposed to autocycles, all scooters and all machines using foreign power units. Not included are the Ariel Arrow series, the BSA Bantam and the Royal Enfield two-strokes, as the first is not of the same class of machine as the content of this book and

Going to work on a Villiers egg. Typical lightweight in town

the others are covered to a greater depth in other Osprey Collector's Library titles.

To save space and avoid repetition the book is divided into Engines, Machines and Appendix. The first deals with the power units in some detail and is arranged by engine make and capacity. The second part covers the makes in alphabetical order with each one taken through its range with notes on colours included. The appendix contains Villiers engine numbers then specifications, first of engines and second of machines, followed by engine and frame numbers, where obtainable, and carburettor settings.

Part One
Engines

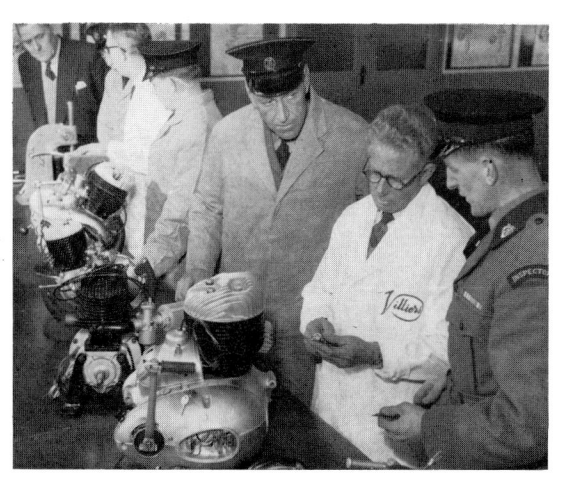

Villiers training school with the AA men receiving their instruction

Villiers

Pre-war

The Villiers two-stroke engine dates back to 1913 and since then a very large number of units for all types of use have been built. In 1960 they passed the $2\frac{1}{2}$ million mark and industrial engines are still in production in the 1980s.

The company was started in Villiers Street in Wolverhampton and later moved to the Marston Road address where they occupied a large site extending over $17\frac{1}{2}$ acres in 1960. One of their great advantages was that thanks to their build-up over the years they carried out nearly all their own manufacture and had pattern shop, foundry, forging drop hammers, coil winding and many machine shops to cover every conceivable method of removing stock.

Thus raw material came into one end of the factory and complete engines went out of the other so that any problems in between could be dealt with in house. Despite the quantity built and the variations in specifications often called for on the same engine for different customers, the production build system allowed for this and ensured each unit left with the correct final drive sprocket and carburettor, the latter also made by Villiers.

In later years this was controlled by using a three digit specification number followed by a single letter for each engine build. This prefix would be stamped on the crankcase followed by a serial engine number and enabled both factory and dealers to keep track of the units. In pre-war days a system using one or more prefix letters followed by a serial number was used.

The first engine, built in 1913, was of 269 cc with bore and stroke equal at 70 mm. It had a one-piece head and barrel with horizontal fins all the way up to the plug. The piston, like the cylinder, was cast in iron and had a deflector top and fixed gudgeon pin. Ignition was by a magneto mounted in front of the crankcase and driven by chain. Lubrication was by drip feed into the crankcase from a hand pump with sight feed in the motorcycle tank.

This was the Mark I and it was followed by Marks II to V between 1913 and 1922. All had the same head finning but late type IV and Mark V engines had the self-contained flywheel magneto used from then on by all Villiers engines well into the postwar period and the advent of coil ignition.

The format of the engine was also settled with the crankcase in aluminium and split vertically with the iron barrel held to it on four short studs. Four bolts held the head in place when it was separate and the built up crankshaft carried the ignition rotor on the right and drive sprocket on the left. In some cases an external flywheel was mounted outboard of the sprocket.

In 1922 a whole range of engines were introduced and the smallest was of 150 cc. This was the Mark VI-C of 55 × 62 mm dimensions with iron piston and fixed head. It was followed by the VII-C and VIII-C, the latter, built in 1924, the first unit to have a fully floating padded gudgeon pin. Lubrication was by petroil and the unit remained in production until 1947. Alongside this is a further 150, the Mark XII-C, was introduced in 1931 with dimensions of 53 × 67 mm and the barrel had twin exhaust ports and a detachable induction manifold. There was also a single port version and both were petroil lubricated, with bronze bush mains, and a cast iron deflector piston fitted with a patented inertia ring whose object was to prevent carbon formation and ring gumming.

Slightly larger were the four 172 cc models of 57·15 × 67 mm dimensions and both sports engines had one-piece barrels, iron pistons and twin exhausts. One was lubricated by petroil but the other had a special automatic system devised by Villiers and used by them for many years. This used crankcase pressure to force oil from a tank, past a regulating screw, into the engine to feed the piston face and via drillways through the crankshaft to the big end. Both the oil feed and the pressure line were via ported holes in the mainshaft so ensuring correct operation. As long as none of the joints leaked air it worked well.

The two other 172 cc engines both had the automatic oiling system; the first was the Super Sports TT. This was fitted with a detachable cylinder head, aluminium piston and a magneto giving variable ignition timing. The second was the Brooklands built in small numbers for racing with high compression cylinder head, padded crankshaft, larger carburettor and a cylinder barrel with shrunk-on, heavily finned aluminium jacket.

Slightly larger still was the 196 cc Mark 1E introduced in 1928 with dimensions of 61 × 67 mm and offered with petroil or automatic oiling. It was a twin port unit but in 1930 it was joined by the single port Mark 2E which used petroil while all three units had iron pistons, inertia rings, detachable inlet manifolds and variable ignition.

A Super Sports 196 was also built from 1929 with twin exhaust ports, alloy head and automatic oiling. In 1938 a new engine of 197 cc was introduced with 59 × 72 mm dimensions, a flat top piston and four transfer ports. It was built in unit with a three-speed gearbox, was called the Mark 3E and became the fore-runner of many, many engines.

In the 250 cc class dimensions of 67 × 70 mm were used from the 1922 Mark VI-A right up to the XVI-A of 1940. Construction followed the lines of the smaller units and the engines had the extra flywheel on the left. The final version had only three head bolts and petroil lubrication although earlier ones had used the automatic oiling.

From 1932 a second series was built using dimensions of 63 × 80 mm and these, the Mark XIV-A models, were available with air or water cooling. Both petroil and automatic oiling were used and the engines had aluminium pistons and plain mains. The Mark XVII-A engine introduced in 1934 used the same dimensions but had ball bearing mains with crankcase compression held by bronze bushes with very close clearance to the shaft. It was followed by the Mark XVIII-A fitted with spring-loaded gland bushes to seal the crankcase.

The 350 cc engines built from 1922 to 1928 used 79 × 70 mm dimensions and ran from Mark VI-B to Mark X-B. In 1931 came the longstroke

The 1939 122 cc 9D unit with three-speed gearbox and twin exhaust ports; built until 1948

70 × 90 mm Mark XIV-B and all these units followed the smaller ones in their specification. In 1931 they were joined by the Midget model, a 98 cc unit of 50 × 50 mm dimensions built for ultra-lightweights. It later became the Junior and then the Junior de Luxe.

Postwar

100 cc F unit After the war the Junior de Luxe engine continued in production up to 1948. It was a simple unit based on the earlier Junior with fixed cylinder head, while the de Luxe had a separate aluminium head and a flat top piston. The crankpin was overhung and the alternative steel and bronze rollers that made up the big end bearing were retained by a large washer held by a rivet in the hollow crankpin.

The crankchamber was sealed with a simple cover on the right and the crankshaft carried the flywheel magneto on its left end. In the centre was fixed a sprocket which drove back to a clutch on a slave shaft carrying the output sprocket. The crankcase left side was extended back to form the right chamber side for the clutch and mated to a left casting that ran forward to support the flywheel stator. The carburettor was on the left, the silencer box underneath and the unit was hung below the bike frame from two lugs.

Late in 1948 two new engines were announced both of very similar construction and with 47 × 57 mm dimensions to give 99 cc capacity. The 1F had a two-speed gearbox, while the 2F was similar to the old Junior with a relay shaft carrying the clutch. The engine layout was a complete redesign with flywheel magneto on the right in the open and crankcase door on the left. This item was, however, altered to carry a ballrace main bearing and the crankshaft extended into it.

The engines were designed to fit into a frame rather than beneath it so had three mounting lugs on the crankcase and a cylinder only tilted forward a little from the vertical. A separate alloy head was fitted and that on the 2F carried a decompressor which was not provided on the two-speed model. The crankcase extended back with a cover on the right to form a primary chaincase, clutch chamber and gearbox on the 1F. The cover also supported the mainshaft with a ballrace but not the layshaft which ran in a bush in the crankcase and one in a bridge bar bolted to it. The layshaft was fitted beneath the mainshaft on which slid a dog clutch to engage either gear with control by a handlebar lever connected to the selector by cable. On the two-speed 1F a kickstarter was fitted on the right and both units used Villiers carburettors.

It was October 1952 before there was a change to the 99 cc engines and then the 1F was replaced by the 4F, although the single-speed 2F ran on unchanged. Like its predecessor, the 4F had a two-speed gearbox and retained the 47 × 57 mm dimensions but within the crankcase was of a considerably altered design. It was

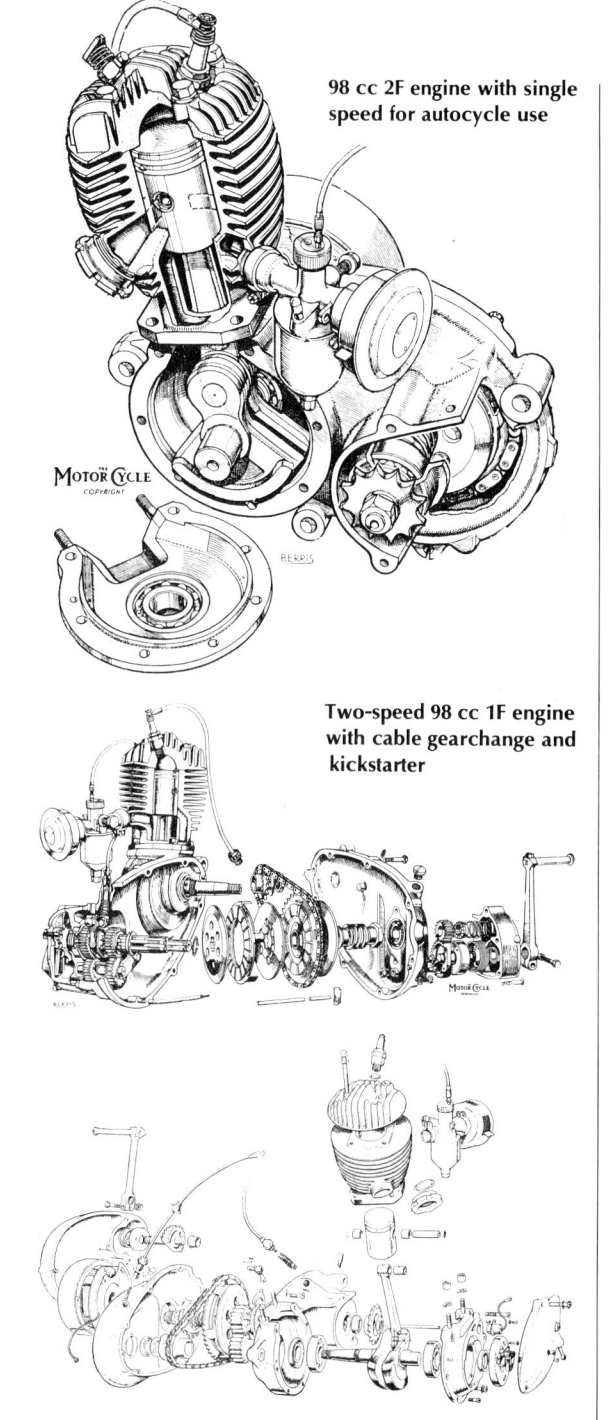

98 cc 2F engine with single speed for autocycle use

Two-speed 98 cc 1F engine with cable gearchange and kickstarter

Later 4F engine with enclosed magneto and remote points

much more streamlined and had an outer cover on the right which enclosed the flywheel magneto and the kickstart mechanism. Unlike the other engines the points were divorced from the magneto onto a plate on the left crankcase chamber cover where they were opened by a cam carried on the crankshaft end. A wire connected the two sections and ran across the engine within a hole in the crankcase.

In 1956 a version with foot control of the gearbox was introduced as the 6F and for 1957 this was made available with hand change. The engine was identical to the 4F but the change was lighter to operate while the gear ratio on all was lowered a little from 1956. A new carburettor, the S12, appeared early in 1957 and, although the three engines were listed for the following year, the demand for the 2F slackened off as mopeds began to appear in large numbers, and so it was dropped in 1958.

The 4F went around the same time but the 6F with its footchange continued on into the early 1960s.

125 cc D unit This series was first introduced about 1937 as the Mark VIII-D with 3-speed handchange gearbox built in unit with the engine. It had ball bearing mains with close running bush seals, twin exhaust ports and a separate cylinder head with plug on the left and decompressor on the right. It was followed in 1938 by the 9D which had some detail changes and during the war this was further altered with the addition of a seal on the drive side.

The 9D followed the lines of the VIII-D and had an iron barrel and alloy head. Three races supported the pressed-up bob-weight crankshaft and the crankcase was split vertically along its centre line. The primary chaincase, on the left, was a two part alloy casting, and the outer one was held by a single nut. A single plate clutch drove the direct top, constant mesh gearbox which was controlled either by a long hand lever or by a short one working in a tank quadrant and connected by a linkage. In either

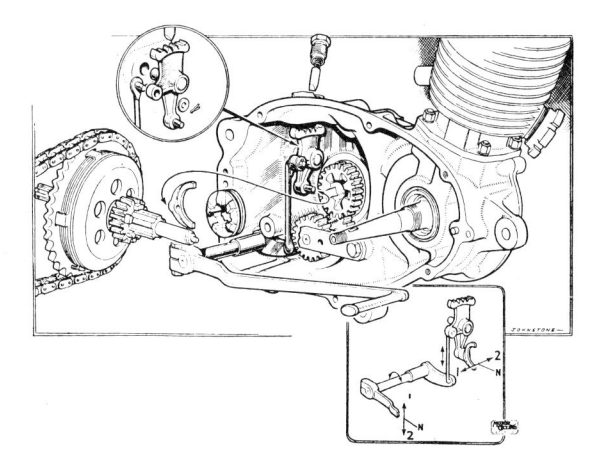

Final 98 cc engine, the 6F with footchange but otherwise as the earlier unit

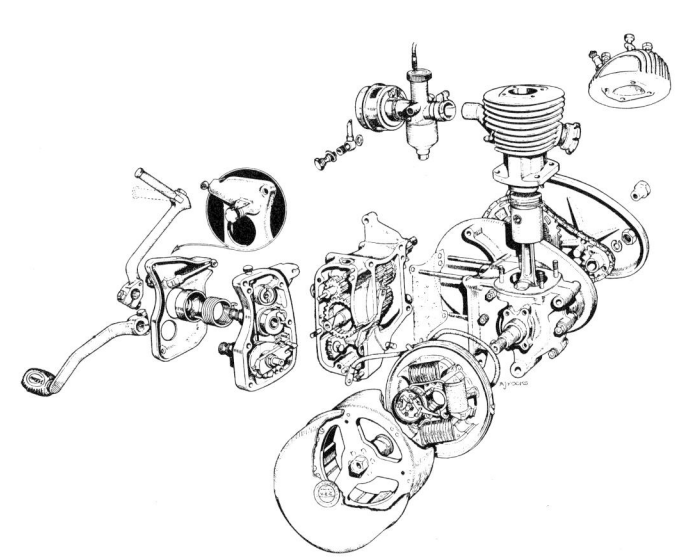

The 122 cc 10D introduced in 1948 with the nearly identical 6E and later versions

case the lever turned a shaft with a helical gear segment that mated with a helical rack cut in the top of the selector fork.

This unit was fitted by James to their military model and saw considerable war service for the machine was light and easy to handle. In that installation the carburettor was slightly different and had a remote choke lever and flap in the

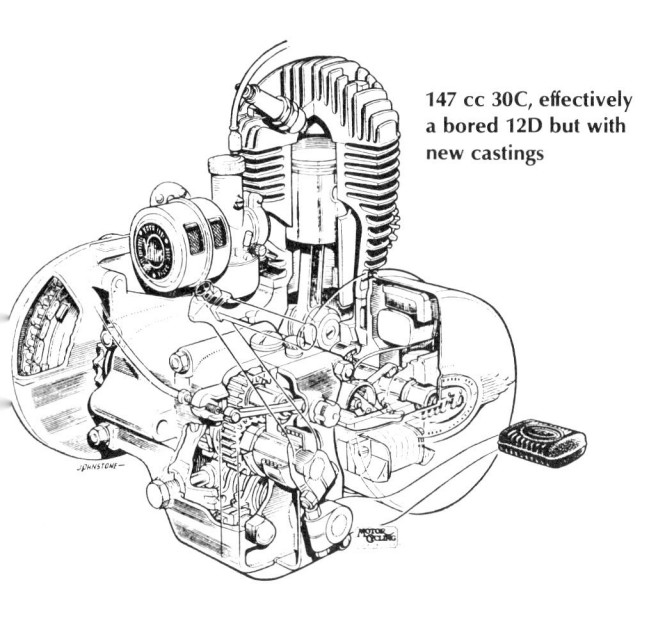

147 cc 30C, effectively a bored 12D but with new castings

Later 31C of 148 cc, shorter stroke and more streamlined

hose connecting it to the air filter.

It was replaced late in 1948 by the 10D which was a new design type to be used by Villiers for a decade. The 10D retained the old dimensions of 50 × 62 mm and was based on a crankcase split vertically on the centre line. Within it the crankshaft turned on three ball bearing mains and was of typical Villiers pressed-together construction with bob-weights and roller big-end.

Crankcase sealing varied and at first both sides used a spring loaded bronze bush, but that on the drive side was changed at the end of 1949 to an oil seal. Such engines have a suffix letter D in their number and from specification 932/1000D a seal was fitted to the magneto side as well.

The barrel was set vertically onto the crankcase to which it was secured by four studs and nuts. An alloy flat top piston with two pegged rings ran in it and it was closed by an alloy head carrying its plug angled back to the rear. The Villiers carburettor was clipped to a detachable inlet manifold and the exhaust port screwed externally for a nut. Ignition and lighting were by flywheel magneto on the right under a polished spun aluminium cover, and the transmission was on the left. A separate chaincase was used with inner and outer halves, the latter carrying a flying wing emblem cast onto the surface from 1950.

The gearbox was contained in a separate casting, open at the front and fitted with an end cover on the right. It fitted onto four long studs in the rear of the crankcase and mated and sealed against a machined face on them. Three speeds were provided and the initial rather wide ratios were brought closer after about a year of production. A kickstarter was fitted on the right along with a very neat clutch adjustor that was fast and easy to use, and low down went a foot pedal for gear changing. The necessary positive stop mechanism fitted within the end cover and this was itself finished with a second, outer cover.

The 10D engine continued virtually without change up to 1953. About 1951 an indication of the variety of builds that was possible was shown by an electric start devised by Invacar for their invalid carriages. This was the start of a large range of options offered in time by Villiers themselves with choices of three- or four-speed gearboxes, reverse gear, fan cooling, electric starting, wide and close ratios and competition specification.

Villiers Singles & Twins

13

Among the builds was a competition 10D with four-speed box used in 1953 but by then the engine was due to be replaced by the 12D which was put into use in the 1954 season. It was distinguished from its predecessor by four horizontal cast lines on the outer chaincase and a number of detail changes, but overall the units were interchangeable. Among the changes were a small closing of the gearbox ratios and improved gearbox oil sealing.

Along with the 12D came a competition version with special cylinder, bigger carburettor, raised compression ratio and special flywheel ignition unit. This was the 11D and was available with trials or scrambles gear ratios in the box. As the 11D/4 it was also available with a four-speed gearbox fitted with wide or close ratios.

One final 125 cc engine was built in 1953, this being the 13D. It was a utility unit and comprised the 12D crankcase with a 10D head and barrel but lacked the normal air cleaner and magneto waterproofing.

All the 125 cc units were dropped in 1954 when their job was taken over by a slightly larger engine.

150 cc C unit After the war the long running Mark 24C continued in production but not for motorcycles as it was fitted to invalid carriages and thus fan-cooled.

For motorcycle use the 30C was introduced in 1954 and to all intents was a bored out 12D. In fact the bore increase meant a larger crankcase mouth aperture and thus new castings, but otherwise the unit followed the construction of the 12D and its three-speed gearbox. Along with the 30C came the 29C/4 with raised compression ratio, four-speed gearbox and improved ignition from the 11D/4. It was later made available with the three-speed unit as the 29C but in either form was dropped by the end of 1956.

In that year the 31C was introduced alongside the 30C but in a very different guise. Engine dimensions were nearly square at 57 × 58 mm and the whole unit was much more streamlined, although still built up on exactly the same lines. Both three- and four-speed versions were available and within the crankcase turned full circle flywheels. The crankshaft ran on two ball bearings on the drive side and a single roller on the magneto which had its cam mounted outboard under a separate small cover. This attached to a streamlined cover that enclosed both the magneto and the kickstarter mechanism.

The head and barrel continued as in the past but tilted forward a little, while the crankcase and gearbox castings ran higher so a hole was provided in them for the float chamber. A drain ran to the underside of the unit where a speedometer drive bolted to the gearbox. The boxes used were common to other models and the 31C was offered for other uses with fan cooling and electric start.

The 30C was dropped in 1959 but the 31C ran on in various builds into the mid-1960s.

175 cc L unit Villiers introduced this unit based on dimensions of 59 × 63·5 mm in September 1956 for the following year's programme. It was available with three- or four-speed gearbox and was very similar to and interchangeable with the 31C, many parts being common. The one feature that allowed the larger engine to be distinguished was the plug position in the right side of the cylinder head.

Like the 31C the 2L was used in various forms and from it was derived the narrower 3L fan cooled unit for scooters. This was fitted with a revised gearchange and selector mechanism in 1961 which in due course went into the rest of the range.

The 2L ran on into the early 1960s but was then dropped, although the 3L remained in use until 1966.

200 cc E unit The pre-war engine of this capacity was the Mark 3E with dimensions of 59 × 72 mm and hand change. It was replaced after the war by the 5E built in the manner of the older model and the smaller 9D with twin

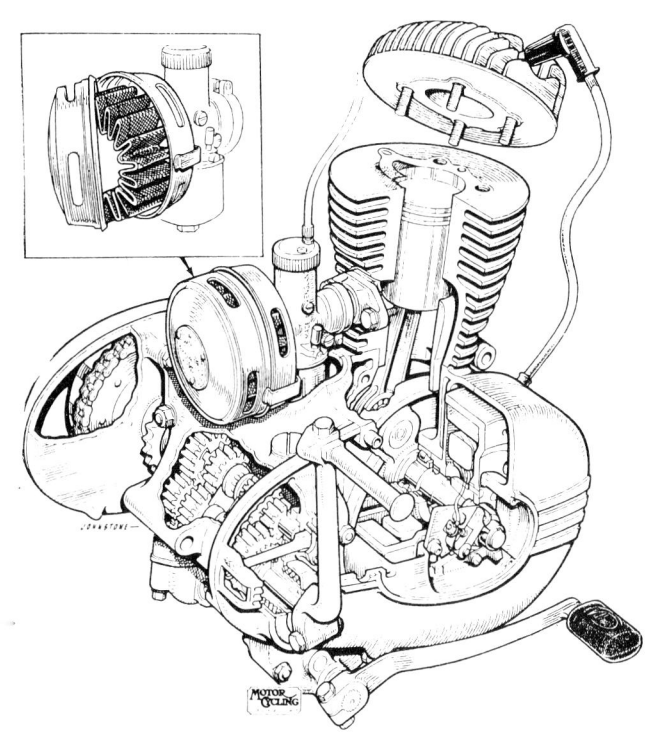

Above **2L of 174 cc with many parts common to 31C. Same construction as most of range**

Below **The four-speed gearbox introduced for both D and E models. Note split bearing for clutch lever**

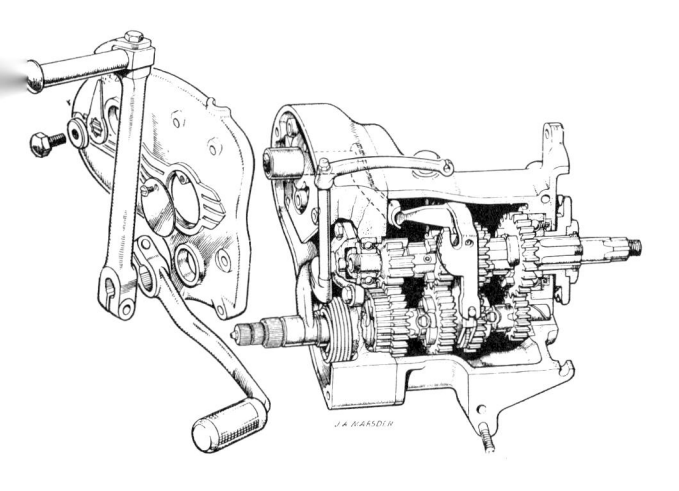

exhaust ports but did have footchange for the three-speed gearbox. It was available from 1946 to 1948 when it was superseded by the 6E introduced late in 1948 and taken up by many motorcycle makers. This unit was of the same design as the 10D but distinguished from it by the cylinder head which had the plug on the right and a decompressor on the left. The crankcase was larger to accommodate the increased stroke but the bearings, transmission design and gearbox were the same. The two units were interchangeable with many common parts, although the primary drive chain size was altered for the 6E.

It underwent the same changes to oil seals and gearbox ratios with two seals being fitted from specification 944/1000D. Like the smaller unit, the outer chaincase was cast with a flying wing emblem from 1950.

The 6E was replaced by the 8E in 1953, the two units being interchangeable and the design followed that of the 12D but with modified engine parts to suit the increased capacity. Detail changes were few but did include the horizontal styling bars on the outer chaincase as on the 12D and the absence of the compression release except when an optional head was fitted. The primary chain size was reduced and a new type, single lever, Villiers carburettor with its own air filter was fitted.

At the same time a competition variant, the 7E, was introduced and this had a high compression head and altered gearing on the same lines as the 11D. Like that model, both 8E and 7E became available with a four-speed gearbox and both three- and four-speed boxes had alternative ratios, the same options applying to the 11D. On the four-speed boxes an index letter was stamped, 'S' indicating standard and 'V' wide ratios. An optional smaller gearbox sprocket was also offered with these boxes.

The 7E was dropped in 1956 but the 8E ran on to 1958 but in three-speed form only for its last two years. In effect both were replaced in 1955 by a new unit, the 9E which followed the same streamlined construction of the 31C. Thus the two crankcase halves were backed onto by a matching gearbox shell and the assembly was hollowed out in the top to clear the float chamber. The flywheel magneto, with remote outboard cam, together with the kickstarter mechanism and gearbox end cover were enclosed by a single outer shell. On it was an access plate to the points.

On the left went the inner and outer primary chain cases enclosing the chain drive to the clutch. On top of the engine the iron barrel was held on four studs and inclined forward a little. The head was in alloy with the plug in the right side at an angle. Internally the crankshaft had bobweights and ran on two ball races on the drive side and a roller on the magneto.

First units were fitted with a four-speed gearbox but soon both three or four speeds with wide or standard ratios were offered. Also available was the 10E which was indentical except that the cylinder was mounted vertically. For 1957 competition versions of the 9E were added and these had an increased compression ratio and improved power. In addition the unit was available with electric starting, fan cooling or reverse gear for other applications.

The 9E continued in use for a good number of years with little alteration but was fitted with the revised gearchange in 1961. It also appeared as the 11E for scooter use and, in its various forms, ran on into the late 1960s.

225/250 cc H unit This, the 1H, was announced late in 1953 as a touring unit and was the first to have streamlined construction. It was based on dimensions of 63 × 72 mm to give 224 cc so was essentially a bored out E unit on the lines of the later 9E, not the then current 8E.

The external design was as for the 9E but carried further to include a cover over the carburettor, which was totally enclosed. Internally, full flywheels turned in four ballraces and the flywheel magneto was much smaller so

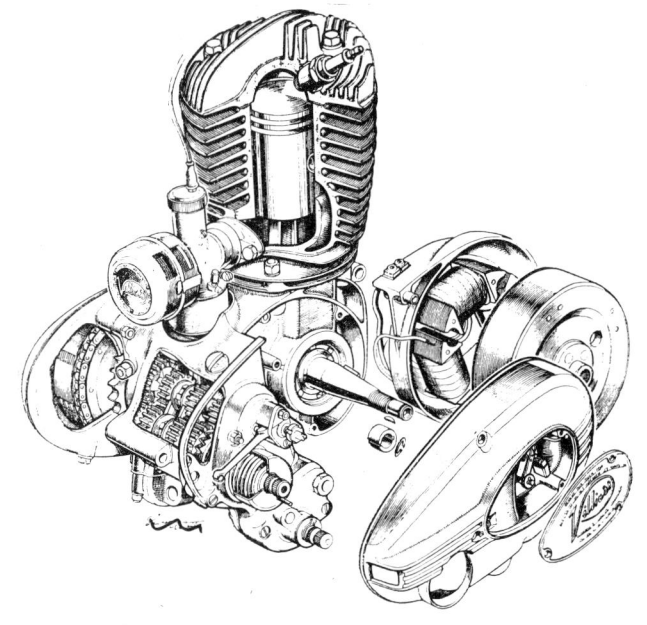

Above **The final 197 cc engine, the 9E. The 10E had a vertical cylinder but was otherwise the same**

Below **225 cc 1H with key in side cover. The first streamlined unit**

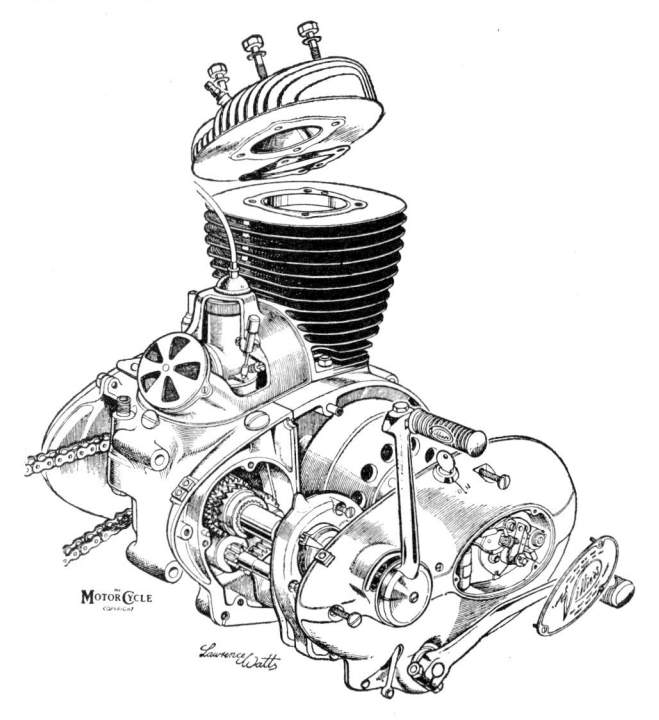

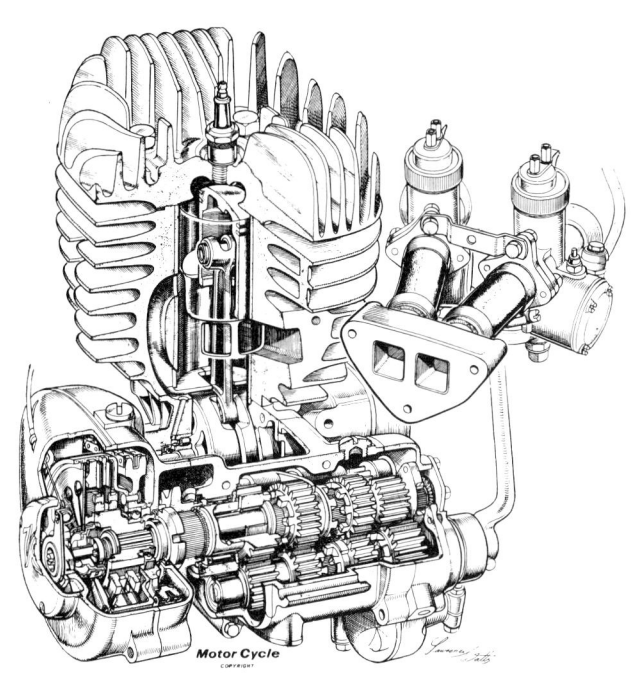

Above **The Starmaker with twin carburettors and special combustion chamber shape**

Below **Final A series, the 37A of 246 cc and based on the 9E with many common parts**

an external coil had to be used. The points remained under a cover on the right and this also carried a key-operated ignition switch in its upper surface.

The remainder of the unit followed Villiers practice and the four-speed gearbox was fitted. Early engines, up to number 2929, were fitted with a primary chain tensioner but this was then removed due to noise.

The 1H was replaced in 1957 by the 2H, basically an enlarged version with dimensions of 66 × 72 mm to give 246 cc and of similar appearance to the 250 cc twin unit. Like the 224 cc model, it was fitted with a four-speed gearbox and ignition switch in the top of the right side cover. It did not remain in production for long as it was superseded by the A range engines.

250 cc A unit The first of these, the 31A appeared in 1958 and in essence was a 9E bored out to 66 mm and 246 cc, so repeating the 2H dimensions. It had a different length rod and was first fitted in the Bond 3-wheeler but for this use was replaced by the 35A late in 1961. The first engines offered for motorcycle use were in trials or, as the 33A, scrambles trim with suitable internal adaptations and were joined and later replaced by the 32A trials and 34A scrambles units introduced in 1960. Many of the parts continued to be common to both the 9E and A range as well as the L and C ones, while the electric start and reverse gear could also be fitted.

In 1962 the 34A became the 36A with a change to the magneto cover that left the kickstarter and clutch exposed for rapid attention, and a smaller chaincase to improve the cooling of the mains on that side. In 1965 the 32A became the 37A for the same reason and both engines continued for a further year or two.

Starmaker Late in 1962 Villiers announced a new 250 cc competition engine unit to be called the Starmaker (actually Star-Maker, at first). It was designed by Bernard Hooper and intended for scrambles use although it was also adopted for road racing and trials.

Part One Engines

The engine followed Villiers practice in so far as the separate gearbox and two-part primary chaincase but broke with their traditions in nearly every other aspect. Bore and stroke were equal at 68 mm and the crankcase was well ribbed externally for rigidity. The crankshaft had full flywheels and ran on two lipped roller bearings close in, and a needle race just behind the engine sprocket on the left. The big-end was a caged needle roller and the rod very slim to assist the high primary ratio. A bushed small-end carried the pin and a two-ring piston.

The barrel had an austenitic iron liner rough machined to ensure a good bond to the cast alloy muff which had well spaced fins, isolated around the inlet and exhaust ports. Later models had the fins pitched more closely. The head was spigoted to the barrel and gave a compression ratio of 12:1 with a wide squish band and central plug. Head and barrel were retained by sleeve nuts on four long studs screwed into the crankcase. An inlet manifold was bolted to the barrel and carried twin Amal Monobloc carburettors which opened up one after the other. A flywheel magneto with external coil went on the right of the engine and transmission was by duplex chain to a diaphragm spring clutch with all metal plates.

The gearbox contained four speeds and was of a new design using needle races throughout, while the dogs were undercut for positive engagement. The original engines produced 25 bhp at 6500 rpm and later Mark II scrambles units reverted to a single Monobloc for simplicity. Road racing engines were fitted with a modified barrel and an Amal GP which pushed the power up to 32 bhp at 8000 rpm.

In time the Villiers company became part of the NVT combine and the Starmaker was used by AJS, another part, to produce road racing and scrambles models in 250 cc and larger capacities.

Twin cylinder units The first of these units, the 2T, was introduced in 1956 and was constructed on the lines of the single cylinder 1H. Its

Above **Later Starmaker with closer pitched fins and GP carburettor**

Below **The 249 cc 2T twin, similar to 9E, A series and 1H units**

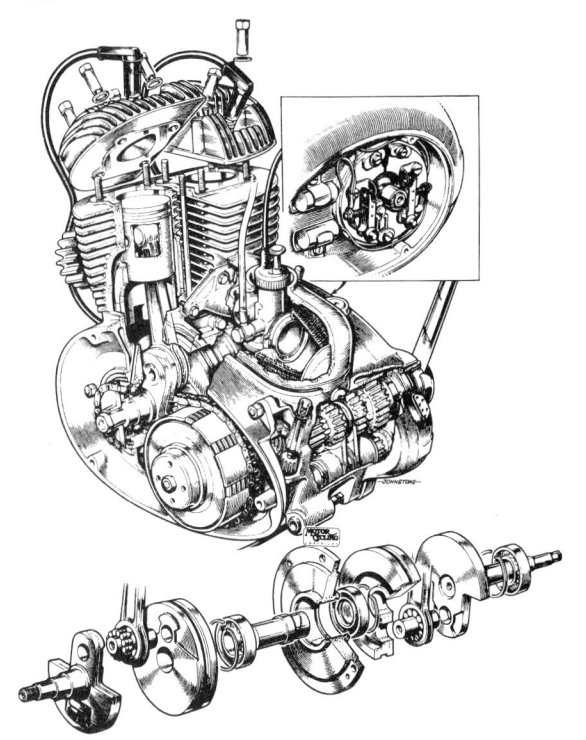

18

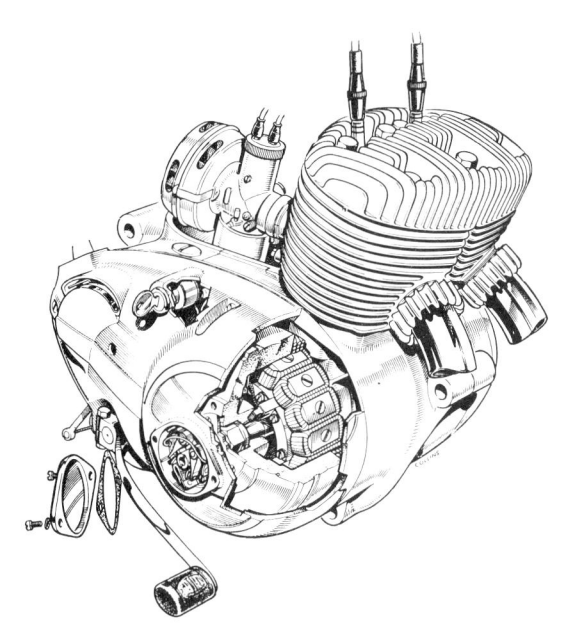

Above **2T fitted with Siba Dynastart. Key controls both ignition and starting**

Below **Cutaway of 4T showing extra transfer ports and typical long Villiers piston**

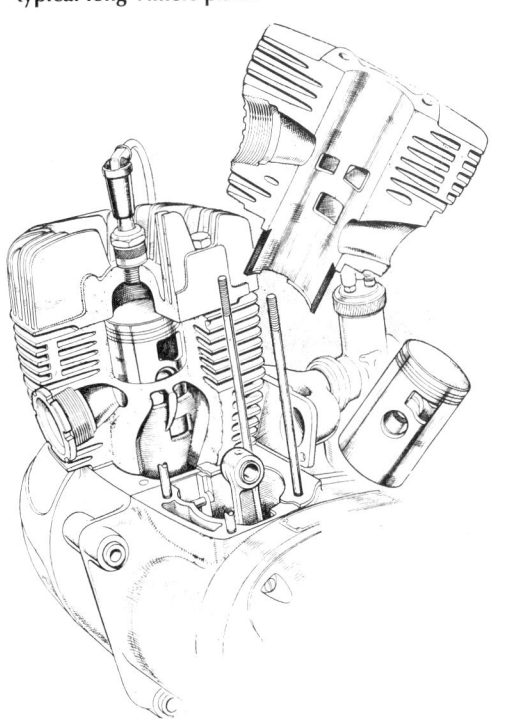

dimensions were 50 × 63·5 mm and it retained the separate four-speed gearbox, ignition key switch and flywheel magneto of the singles. While the fixing dimensions remained the same, the twin was fitted with rubber bushes in the lugs to combat vibration and a central disc was bolted in the crankcase to separate the two chambers. To do this it carried an oil seal in addition to a roller bearing duplicated behind the magneto and assisted by a ballrace behind the engine sprocket.

The two crankcase halves supported separate iron barrels and alloy heads with a single carburettor under a cover feeding both. The points cover on the right was duplicated on the left and these were often made to carry the machine maker's logo rather than that of Villiers. In use they sometimes finished up with different ones on each side to confuse second or third owners.

In time the 2T was joined by the 324 cc 3T introduced at first for light car use. In 1958 it became available for motorcycle use without the fan cooling and reverse gear which were options for either twin. The reverse was obtained by running the engine backwards, a feature that the use of a special Siba generator allowed, this item also being wired to act as an electric starter on demand. The 3T was essentially just a 2T bored out but after the early production the engine sprocket was made larger.

In 1961 the twins received the same gearbox improvement as the singles and in 1963 the 2T was replaced by the very similar 4T except for one user who continued with the older unit until 1968. The 3T was discontinued in 1964

The 4T was little changed from the 2T but the barrels had four transfer ports not two, ported pistons were used and the key switch in the right cover was deleted. It was built up to 1967 in various forms.

Square barrels This phrase referred to kits which became available in the late 1950s, at first for the 9E unit and later for the A series. In the first case they usually enlarged the engine to 250 cc and in all comprised a special barrel, piston, head and four long studs to hold the items to the standard crankcase. The barrels were cast in alloy either with an iron liner or would have a hard chromed bore. The fins tended to be well spaced and of a square layout, hence the name.

Firms involved included Parkinson, Marcelle, Vale-Onslow and Greeves, while in time the last also made use of the Alpha lower half to produce their own engine.

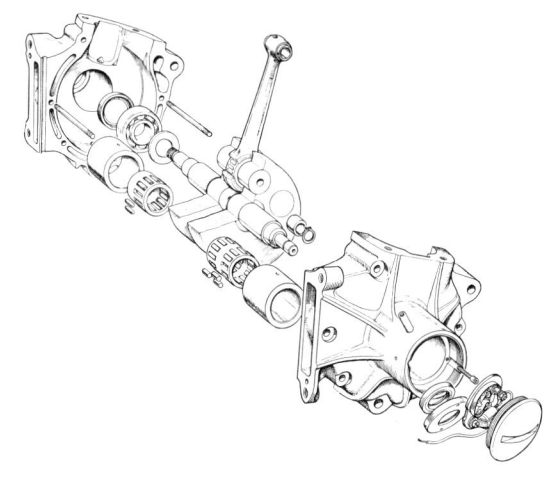

Above **Alpha bottom half with full flywheels and stronger cases**

Below **A square barrel conversion, in this case a DMW on a 34A engine in their own machine**

Alpha

Alpha Bearings of Dudley used to and in 1982 still do make replacement big-end bearings for motorcycle and other engines. The firm came into being in a small way in 1946 to supply an item that no-one else did and as is common in such cases, it expanded.

When the square barrel conversions for Villiers appeared so did problems, for the 9E and A bottom halves were not designed for the power they were then expected to handle. The results were that the cranks spread at the pin, the webs then machined into the crankcase, the crank flexing worked the mains loose in their housings, and lots of spares were bought.

The first step for Alpha was a better crankshaft with full circle flywheels and bigger pin but within the constraints of the existing case it was impossible to stop the flexing. The external Villiers flywheel did not help. Step two was a new bottom half which would accept either the Villiers top end or the conversions and be strong enough to cope with either.

The result was an assembly with the same barrel mounts, engine lugs and gearbox fixings as the Villiers but wider with the crankshaft running on two rows of rollers on the right and a combination of ball race for location and rollers on the left, drive side. Construction was conventional but the ignition flywheel was replaced by cam and points for an external battery and coil set-up.

One thing leads to another and, while the factory concentrated on the new bottom half, Frank Cutler the managing director turned his attentions to a rotary valve single using the flywheels as the valve by cutting away the flange which ran very close to the cases. The first prototype had the inlet tract taken into the right side of the crankcase just in front of the shaft with a steep downdraught angle for the Dell'Orto carburettor. Number two had the carburettor behind the barrel on an inclined inlet tract leading to the back of the flywheels. That engine had a third, rear transfer port which communicated to the under piston crown area. It also had a positive oil supply to the main bearings, although it retained its petroil, albeit with a leaner ratio. That unit went into Royal Enfield cycle parts and went well when revved.

The single may have performed nicely on the road but it refused to rev high enough for racing, so Frank Cutler built a twin. This had dimensions of 54 × 54 mm so was of 247 cc capacity with the two parallel cylinders inclined forward a little. Disc valve induction was used but because of the need to mate with a standard five-speed Albion gearbox the layout was unusual.

Only one carburettor was used and this was mounted, like the second single, on the engine centre line behind the crankcase. It fed into a tract that split into two and the discs were mounted on the inboard side of the flywheels, keyed to the mainshaft, and either side of the centre ballrace main bearing. Roller bearings supported the crankshaft between the disc and the inner flywheel with needle races at the outer ends.

Internally the rest of the engine was conventional and coupled to the gearbox with a duplex chain. Ignition was by coil with the points on the right and the whole unit was mounted in Royal Enfield GP cycle parts and called the Alpha Centuri.

The engine was built in 1965 and raced for the next two years with minor modifications to improve performance. The cycle parts were DMW for a time but finally became Alpha frame and Ceriani forks. In that form the model continued to be raced for a few more years but ultimately had to give best to Yamaha.

Top right **First Alpha single with side carburettor and rotary valve**

Right **Later Alpha single with rear carburettor and well finned head and barrel**

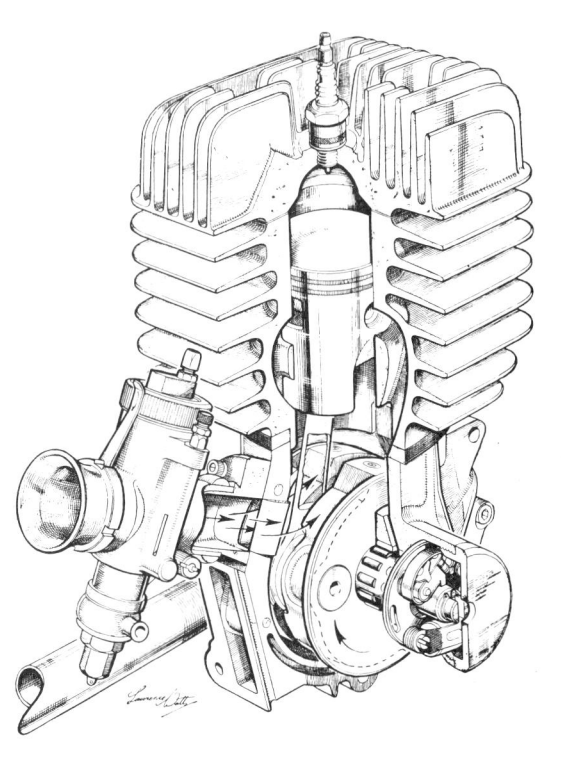

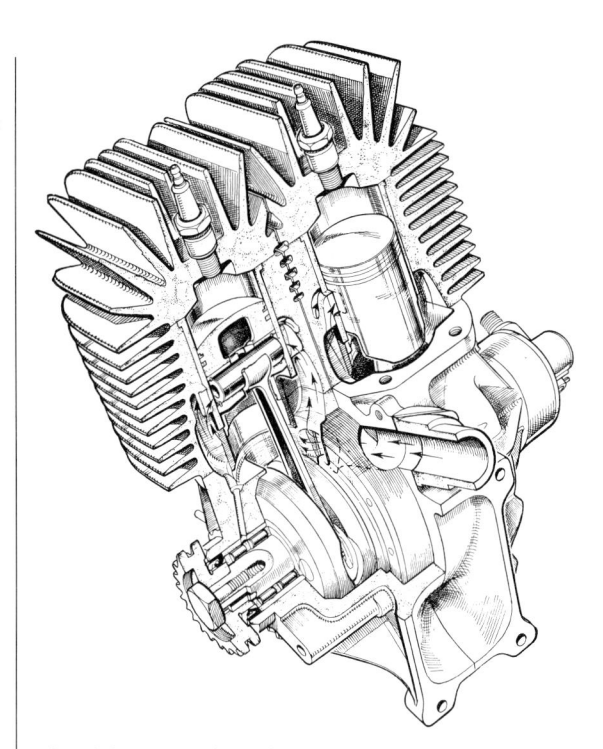

The Alpha twin with single inlet port and twin disc valves

AMC

The first appearance of AMC engines occurred in 1953 when DMW announced that for some of their range for the next year they would be using units supplied by Ateliers de Mecanique du Centre of Clermont Ferrand.

These were two in number, a 175 cc ohv and a 250 cc ohc, both being single cylinder four-strokes with unit construction gearboxes.

Three years later the completely unrelated English AMC group based at Woolwich announced their own single cylinder two-stroke engine and gearbox unit to be used by group members Francis-Barnett and James. There was some confusion.

The (English) AMC unit was a very well streamlined one similar to the Villiers 1H with the carburettor enclosed but contained several novel features. Internally most of the engine followed convention with full flywheels, caged roller big-end, three main bearings, iron barrel and alloy head, although the last did have radial fins. Unusual were the transfer ports which were without the normal inner walls so were just depressions in the bore. To guide the mixture the piston carried transfer ports, its crown rose steeply at the sides well above piston ring level, and the head had a pair of deflectors pointing down and matching the cutaways in the crown and sides of the piston.

Dimensions were 66 × 72·8 mm to give 249 cc, compression ratio was 8.25:1 and the engine drove to a four-speed gearbox via a single strand chain on the left. The box was made by AMC and fitted into its own shell which blended with the crankcase. A single outer cover on the right enclosed alternator and gear mechanism with the points under a further small cover on it.

This engine was followed by a 171 cc version in 1957 which used dimensions of 59 × 62·7 mm, while in 1959 one of 199 cc appeared using a dimension from each to produce 59 × 72·8 mm and thus a small bore version of the 250.

The same year also saw a smaller engine still of new design using the 171 stroke but a 55 mm bore to give 149 cc capacity. The new engine followed convention much more internally but was contrived so that the parts could be assembled to give a cylinder position in the vertical, horizontal or inclined. This was done by providing lugs on the three-speed gearbox shell so the whole crankcase could be rotated to fit in the desired position.

Inside were bobweights, two-ring piston and normal practice except that a second set of transfer cut-outs were cast into the crankcase mouth to allow the barrel to be turned if this was needed for installation.

A pressed steel chaincase was used with the inner part helping to support the gearbox, while on the right a pressing enclosed the flywheel

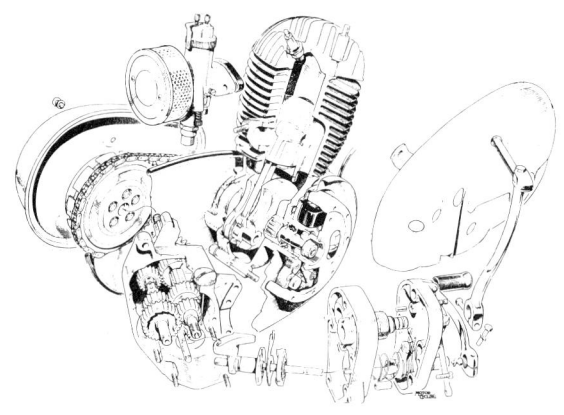

149 cc AMC single with three cylinder positions available on assembly

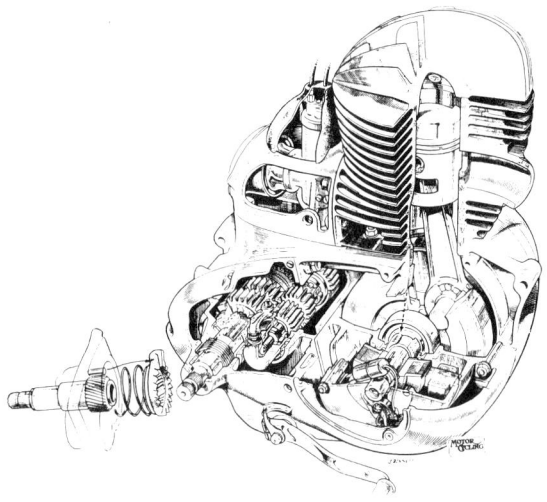

Above **First AMC single of 249 cc with unusual piston crown and open transfers**

Below **This, the 322 cc Anzani twin, had piston and rotary inlets**

magneto and gearbox end. An Amal carburettor was fitted.

In 1961 AMC announced that they were to cease manufacture of their range of engines and the initial effect was for assembly to be carried out by Villiers at Wolverhampton while the group worked with Francis-Barnett and James on new designs.

In practice these never reached production and the AMC units were used for varying periods with the 171 going from the range in 1960, the 249 in 1963, while the others remained until 1966 when production of them ceased.

Anzani

Anzani is a very old name in the motorcycle industry and, as a French company, dates back to Edwardian days. British Anzani came into being after World War 1 and were associated with big vee twins, often with eight valves, and marine engines.

In 1953 they announced two new boat engines, both parallel twin two-strokes with a common crankshaft of 57 mm stroke. Coupled with a 52 mm bore this gave the smaller a capacity of 242 cc while the larger was of 60 mm bore and 322 cc.

The crankcase was split horizontally and the engine watercooled but its most unusual feature lay with the rotary inlet port. This was combined with the large central plain main bearing which had a split bush. The crankshaft journal had two angled holes in it, each running from the bearing surface to the inside face of the inner flywheel where it emerged just inboard of the crankpin.

The centre bearing was surrounded by an annular passage within the crankcase casting that was ported inwards to a single port in one half of the bearing shell and to the induction stub. As the crank rotated the intake was thus connected to each chamber in turn.

Late in the year a motorcycle version was announced in the smaller capacity and fitted with an air-cooled head and barrel. The latter carried a stub fitting Amal and had slightly splayed exhaust ports. A Wico-Pacy generator was fitted on the right and a duplex chain drove a four-speed Albion gearbox.

In 1954 the enlarged 322 cc version was listed and, although very similar to the earlier engine, it differed in that it had piston controlled inlet ports in addition to the rotory valve. In 1955 this unit was fitted with a Siba generator which doubled as a starter, and had a reverse gear to make it suitable for three-wheeler use. For 1956 a water-cooled 242 cc unit was produced and fin area increased on the air-cooled ones.

The engines remained in production for some years but their use in motorcycles ceased in 1960.

Excelsior engines

The Excelsior engine manufacturing shops came into being during the war to produce units for the Welbike and thus were available and ready to go into civilian production in 1946.

In this way they came to be one of the few firms not dependent on Villiers, although they used this make for many of their models as they had in pre-war days.

The Welbike engine was a simple single cylinder 98 cc two-stroke that was continued in production for the Corgi as the civilian machine was called. For that purpose manufacture was by Brockhouse Engineering of Southport under licence from Excelsior, and the engine was called the Spryt.

The power unit with single speed and clutch was based on two alloy castings which enclosed the transmission and carried the two crankshaft ball races. Onto the front went the horizontal cylinder in iron with alloy head. Engine dimensions were 50 × 50 mm and the barrel was fitted with alloy deflectors in the transfer ports which were fed via holes in the piston. The carburettor was stub mounted on the left and the exhaust connected underneath.

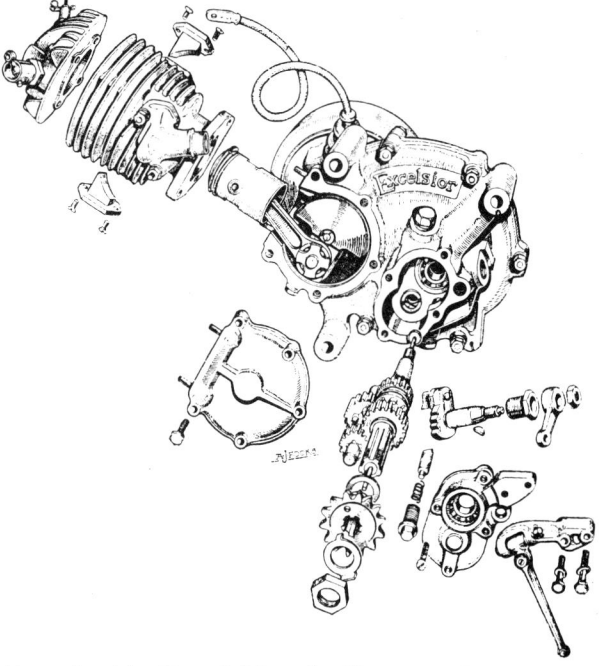

Above **Excelsior 98 cc Goblin unit with two speeds and detachable transfer windows**

Below **Excelsior Talisman Twin unit, 243 cc, four-speed; also built in 328 cc form and as a triple**

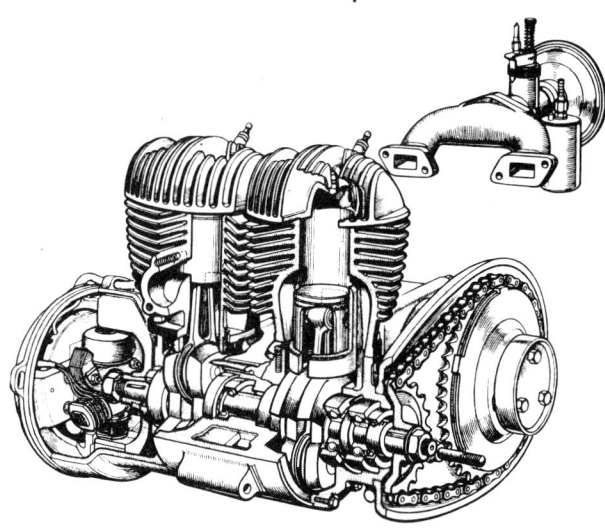

An overhung crankshaft with roller big end was used with the crankcase machined into the outside of the right casting and closed with a simple cap. The left end of the crankshaft carried the flywheel magneto and generator for the lights, and between the mains went a sprocket. This drove a single plate clutch on a countershaft carrying the final drive sprocket and the clutch lever on the left.

This unit was only used by the Corgi up to 1948 when it was taken over by Brockhouse. Before then, for 1947, Excelsior had a new two-speed 98 cc unit designed and built for their own use. The new engine was called the Goblin and reversed the Spryt design with magneto on the right and door for the overhung crank on the left. In other respects it had many similarities but with the cylinder inclined and a more substantial main casting. The transmission was enclosed by a cover which retained the second main bearing but no longer supported the clutch shaft. The major casting had a section on the left for the gearbox which had cross-over drive, direct top and layshaft running in bushes. A simple lever moved the internal meshing dog to select the two gears and neutral. The final drive and the clutch lever remained on the left.

During 1947 a Mark II Spryt engine appeared on the lines of the Goblin but with just a single speed and countershaft clutch. Although the engine was the same as the Goblin the main casting differed as it did not have the gearbox housing, just a boss to carry the ball races for the countershaft. This now had the clutch hung on one end and the output sprocket on the other. It retained the magneto on the right and overhung crankshaft design.

During 1948 the 98 cc Goblin was joined by a 123 cc version, still with two speeds, for use in very lightweight motorcycles. It used nearly all of the smaller unit and achieved the extra capacity with a 56 mm bore. Both engines were fitted with a kickstarter to assist their motorcycle use but were only used until late 1949.

At that point they left the engine range, although the Spryt and Goblin units continued for use in the Autobyks. They were joined by a completely new engine, the Talisman Twin, which was of 243 cc and based on 50 × 62 mm dimensions. The crankcase was in three vertically divided parts so that each vertical cylinder sat on a case joint just as in a single. The outer castings thus acted as side covers to each crankcase chamber but also carried main bearings and seals. There were six bearings in all, a single ballrace on the right side behind the flywheel magneto, two ballraces in the centre with two seals between them, and two more ballraces plus a plain outrigger bush on the left drive side.

There were two crankshafts which fitted together in the centre, keyed to the required 180 degrees between throws. One inner mainshaft was hollow and the second extended through it with a threaded end. This was secured by a special nut which fitted between the bobweights to pull the assembly together. To enable assembly to take place the crankpin on that side was only a light press fit in one web so that part could be added to the engine after the two crankshafts were together.

For the rest the engine was conventional with separate cast-iron cylinders and alloy heads. The gearbox was a four-speed Albion bolted to the back of the centre crankcase and driven by a single strand chain on the left within an alloy case. Final drive was also on the left.

For 1952 the main bearing behind the magneto was changed to a roller race and the primary chain to a duplex type. A Sports Twin was also built fitted with twin carburettors but retaining all the other features.

In October, with the announcement of the 1953 range, came a new single cylinder engine of 147 cc. This was based on dimensions of 55 × 62 mm and was very conventional with vertically split crankcase, three ballrace main bearings of which two went on the drive side, double

25

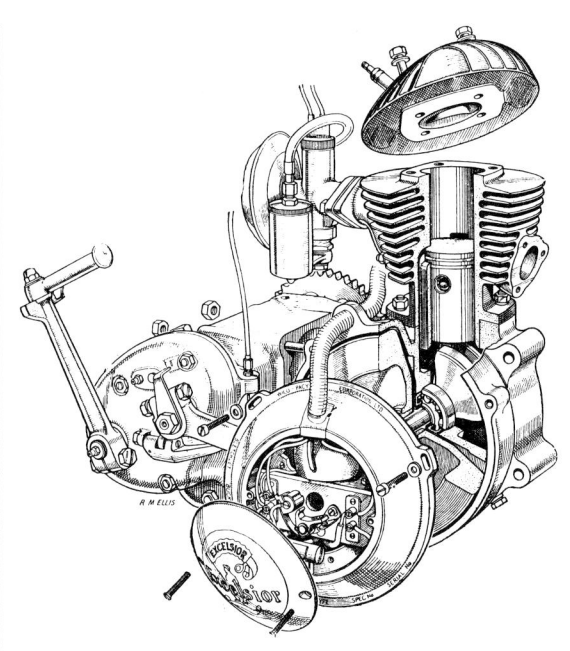

Above **Excelsior single of 147 cc with three-speed Burman gearbox. Very similar to Villiers unit**

Below **The JAP 125 cc unit. Curious helical gear on selector fork common with 9D**

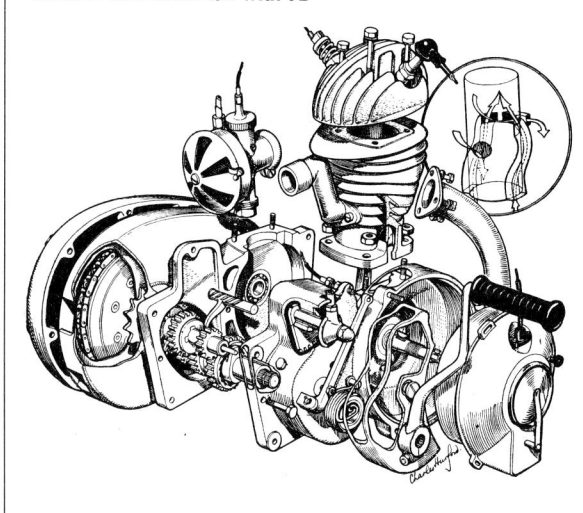

roller big-end as in the twins, iron barrel held on four short studs, and alloy head bolted down to give a 7·8:1 compression ratio. The primary drive was by single strand chain to a two-plate clutch and the three-speed footchange Burman model R gearbox bolted to the back of the crankcase. Ignition was by a Wipac flywheel magneto on the right and an Amal carburettor was flange mounted to the alloy inlet tract and fitted with a pancake filter.

Late in 1956 the Spryt and Goblin engines were dropped as the Excelsior Autobyks which used them went from the range, while the original type Spryt design as built by Brockhouse for use in the Corgi had gone with that model late in 1954.

The Sports Twin was altered a little for 1957 with redesigned heads and barrels with increased fin depth and area. The exhaust ports were modified as was the polished primary drive case. During that year the 147 cc engine was dropped for motorcycle use but continued in production for the Excelsior scooters so only the two 243 cc twins were left. However, they were joined late in the year by an enlarged version produced by boring the cylinders out to 58 mm to give a capacity of 328 cc. Twin Amals were fitted but otherwise the engine was a copy of the smaller unit in design and construction.

A number of three wheel cars took up the 328 cc Excelsior engine and for those seeking a little more power they built a three cylinder version using the same engine dimensions to give 491 cc. In this case a Siba dynastart was fitted and a gearbox with four speeds and reverse.

The twins continued in production with little change until late in 1962 when they were dropped along with the 150 which had made a brief reappearance in a motorcycle that year.

The line of Excelsior engines had come to an end.

JAP

The firm of J. A. Prestwich of Tottenham are best known in the motorcycle world for their speedway engines and big vee-twins. They did in fact make all manner of engines in the early postwar period and one of these was a small, single cylinder two-stroke of unit construction with three-speed handchange gearbox.

It was a conventional unit with dimensions nearly square at 54·2 × 54 mm giving a capacity of 125 cc. The crankcase split vertically on the centre line and carried a pressed up crankshaft with bobweights and a roller big end. It ran on three ballrace mains and the chamber was sealed by bronze bushes, while the four stud iron barrel was capped by an alloy head with plug on the right and decompressor on the left.

Most unusual was the fitting of a cast iron piston which was used to allow a closer running fit. It did have a flat top and the combination of these factors and the twin transfer ports resulted in 3·5 bhp at 4000 rpm. A Wico-Pacy flywheel magneto was fitted on the right, and a single strand chain on the other side took the power to a six-spring clutch.

The gearbox was the same as that used in the Villiers 9D with the helical gear connection between selector fork and hand lever shaft. A kickstarter pedal was fitted on the right and a cover enclosed its mechanism and the clutch lever.

The primary chaincase was also used to enclose the rear drive sprocket and the inner wall was cast with the left crankcase half. Two holes at the rear provided passage for the rear chain. To this casting was attached a flat steel sheet to form the chaincase inner, after which the primary drive was assembled and enclosed by an outer casting held in place by a number of screws.

The engine was fitted by a number of the smaller makers, and for 1951 gained footchange for the gears. In this form it was built until 1952, but was then dropped.

RCA

This was the name given to a machine built in 1957 by R. Christoforides and Associates of Harlesden, London, using their own engine unit in Greeves cycle parts. The engine was designed by Peter Hogan who, with his brother John, had been at the forefront of English 125 cc racing in the early 1950s using a pair of Bantams. John won more races than he lost in that era and later went on to EMC-Puch and then MV lightweights with a spell with an Anzani twin. Peter also had a 250 cc twin but built up from two coupled Bantam engines.

The result of this experience was the 349 cc RCA twin with dimensions of 63 × 56 mm and compression ratio of 8.0:1. The engine was very modern in concept with horizontally split crankcase, full flywheels and electric starting by Siba unit. Twin Amal carburettors were fitted after bench tests with a single Zenith and the exhausts were on the outer sides of the iron block. They connected to alloy adaptors bolted to the block with exhaust pipes running from these down and back to the silencers. Separate alloy heads were fitted as the Hogans had sold such items for the Bantam at one time.

A four-speed Albion gearbox was bolted to the back of the crankcase and the primary chain was contained in an alloy case. The unit was carried in a Greeves frame with its alloy front member and the Greeves leading link forks provided front wheel suspension. At the rear a swinging fork carried the wheel. For the rest the cycle parts were adapted to suit and complete the prototype which was tested by Vic Willoughby in 1958. He was timed at over 75 mph and found that the engine had been tuned for low down pull rather than high speed revs. It made for pleasant riding with a top engine speed of 6000 rpm and good acceleration. The engine was also fitted into Dot frames for a short spell. The company built three machines for road racing with reversed exhaust manifolds and short pipes.

The engine was an enterprising effort.

Part Two
Machines

Aberdale

This was an autocycle launched in 1947 and powered by a Villiers Junior de Luxe engine. It was typical of its type with the engine unit slung from the underside of the frame which had a dropped top tube to assist mounting. A petrol tank was fitted into the space formed by top, down and seat tubes, while the power unit was shrouded by side shields which detached to give access.

The bottom bracket carried the pedalling gear with its own chainguard on the right matched by one on the left to protect the drive chain. The frame was rigid with blade girder forks and the wheels and tyres a halfway stage between cycle and motorcycle sizes but fitted with drum brakes. The front wheel also drove a speedometer.

Equipment included lights powered by the engine generator, bulb horn, rear carrier and stand. Controls comprised a throttle lever on the right, a clutch lever with catch to hold it in the disengaged position, a decompressor and inverted levers for front and rear brakes.

The fuel tank held $1\frac{1}{2}$ gallons of petroil mixed at 16:1, and the performance was typical of the class with a top speed of 30 mph and consumption figures around 150 mpg being achieved.

The machine went into production in the subsidiary Bown factory in Wales, and continued under the Aberdale name to the end of 1949. From then on it was sold as the Bown and is described under that name.

ABJ

During 1949 the company of A. B. Jackson of Birmingham entered the lists of machine makers by making use of the then new Villiers F range of 98 cc units. They built two machines of nearly identical appearance, one a motorcycle and the other, fitted with pedalling gear, an autocycle.

Both used essentially the same cradle frame with rigid rear and telescopic forks which incorporated split cones to provide some damping on extreme fork movement. Engines were the 1F for the motorcycle and 2F for the autocycle, which had the pedals to the rear of the gear casing.

Small drum brakes were fitted in the spoked wheels and the front one was covered by a valanced, sprung mudguard. The rear guard was narrower and two of its stays were joined to form a luggage grid. A single saddle was fitted to a pillar tube so could be adjusted for height. Finish was in black with gold tank lines, while the wheel rims were plated. The motorcycle was fitted with an electric horn but its single-speed cousin had a bulb type, while both had rear wheel stands and very long tailpipe extensions behind their silencers.

In 1952 the company also began to market a cyclemotor and at the end of the year dropped the two 98 cc models from their range.

1949 ABJ motorcycle fitted with 2F engine. Autocycle very similar but with pedals

Colours

Black with gold lining to tank, oval tank badge, chrome plated wheel rims, headlamp rim, exhaust sytem and bars.

AJS

The two-stroke era of the AJS story came late in their day after the group had been acquired by Manganese Bronze and became a section of Norton Villiers. In the upheaval that this brought about the decision to use the AJS name for a 250 cc two-stroke came in part from the performance of the Villiers Starmaker Special in the 1966 TT. This used the Wolverhampton engine in a modified Bultaco chassis and was prepared and ridden by Peter Inchley into third place in the 250 cc event.

For 1967 a new frame was built and with the 1966 engine proved quite competitive. With this engine experience a scrambler was prepared late in the year and this developed into a successful machine during 1968. The result in 1969 was the production Y4 scrambler which had an engine developed from the original Starmaker fitted into a frame with massive tubular backbone and twin engine loops. AJS front forks and a Girling controlled swinging rear fork looked after the suspension and the machine equipment suited the purpose it was built for.

With it came the 37A-T trials model which used the 37A engine unit in a similar frame with a single down tube. In this instance Metal Profile forks were fitted, although Girlings still dealt with the rear suspension. General equipment was to trials specification.

The appearance of these two models gave credibility to the new company for earlier announcements of the Alamos scrambler and Double T road racer, both using Starmaker engines, had not resulted in any machines.

1970 brought a new and larger scrambles machine, the Y5 and its power meant some detail chassis changes which were also incorporated into the Y4. The engine units had by now lost their Villiers Starmaker ancestry and were known as AJS Stormer units. The 250 continued with the 68 × 68 mm dimensions of the older unit but the larger machine used an 83 mm bore to give 368 cc capacity. Otherwise it was the same machine but with a lot more power and virtually no more weight.

Development continued so that in 1971 the Stormer 410 appeared with the same piston size as the 370 but a slightly longer 74 mm stroke which raised the capacity to 400 cc. The chassis was the same as that used by the 250, well finished and very practical.

The AJS road racer fitted with Starmaker engine in 1967

1982 FB-AJS 250 trail model, similar to enduro and motocross machines

The small company managed to survive the troubles suffered by the Norton Villiers group and eventually finished in a small building near Andover in Hampshire, close to other parts of what became NVT. The name was changed to FB-AJS and in the middle to late 1970s the range of models changed to a 250 cc trail and enduro bike plus a pair of moto-cross machines of 250 and 360 cc capacity. All were based on the developed 68 mm stroke engine with its wide or close ratio gearbox. Bore sizes were 68 or 83 mm as in the past and much of the chassis design was common, although the moto-cross models had a very large 12 in. of movement in their ultra-light front forks, while the trail bike had to manage with a mere 9 in.

So AJS continued and built a small number of good machines with a very good spares back up. By concentrating on low down and mid-range power the four-speed gearbox was no handicap and for the average clubman the result a machine that was easy and enjoyable to ride. In 1981 a 360 enduro model was added so that for 1982 the company has trail and moto-cross machines available in two capacities with virtually identical engine units.

A successful arrangement.

AJW

This company took the initials of its founder, A. J. Wheaton, who first built motorcycles in 1926. After the war the name was acquired by J. O. Ball who continued with a range that included a 500 powered by a side valve vertical twin JAP engine, speedway machines, both solo and sidecar plus prototypes.

Two of these used the same layout but very different JAP engines, one a 500 cc single and the other the 125 cc two-stroke. In both cases the cylinder lay horizontal so the gearbox sat above the crankcase and the unit was fitted into a spine frame. This had swinging fork rear suspension

The 1952 AJW lightweight fitted with 125 cc JAP engine and neat tail

with a very high pivot point and telescopic front forks. Large front and smaller rear drum brakes were used and the seat was in dualseat form on a tail unit that ran back to the number plate. Quite an innovation for a 1952 lightweight.

It was a nicely finished prototype but no more was heard of it, although the firm did revive its business in 1958 with a range of imported machines.

Ambassador

The first machine to carry this name was announced in January 1947 and came from works in Ascot run by the pre-war record breaker Kaye Don. His American car import business had to turn to vehicle repair during the war and so was well placed to make a simple lightweight after it.

The machine used the Villiers 5E engine carried in a rigid, brazed frame with two duplex plates running between down and seat tubes to brace them and support the power unit and its three-speed foot change gearbox. The forks were pressed steel Webb with hand controlled friction dampers, while comfort was looked after by a Lycett saddle.

Both wheels carried 3·00 × 19 in. Dunlop tyres on black rims with stainless or rustless steel spokes. Both brake drums were 5 in. diameter and the machine was well equipped for its day with twin exhaust pipes and silencers, direct lighting, bulb horn, toolbox on the right, centre and front stands, and not much ground clearance. Tank finish was in silver lined black and red.

In 1948 this model, the Series I, became the Series II with a welded tubular frame and then for 1949 the Series III with the Villiers 6E engine. This ran on to 1951 with a battery and rectified alternator output, but for 1950 was joined by two new models. The Series IV, also called the Popular, retained the original direct lighting, while the Series V broke new ground with the fitting of telescopic front forks.

For 1951 there was a new model and a change of names and colours for the existing ones, although the power unit remained the faithful 6E in all cases.

The new model was the Supreme which was fitted with plunger rear suspension and MP telescopic front forks. Its front mudguard was a little unusual as, although it was supported with front and rear stays to the lower fork, its other location was a bracket bolted to the top of the guard with oil lubricated, felt lined, semi-circular ends which slid against the fixed fork tubes.

The finish of the Supreme was in a new stove enamel with greater durability and the basic colour was grey. This was carried through to just about every detail including the battery and footrest rubbers, while the tank was chrome plated with grey lined panels.

On the other models the Series III continued as the Courier with the tank finished in the new grey finish, the Series V became the Embassy and changed its tank colour and its telescopic forks to MP, while the Popular carried on with its black rims and no kneegrips but did get the grey tank finish.

The Courier was dropped at the end of 1951 but the other three continued and were joined in October 1952 by a sidecar outfit. This was sold only as a complete unit and comprised an Embassy model fitted with special Webb girder forks and coupled to a single seat sports sidecar finished in black with red upholstery.

The Supreme was altered a little and fitted with wheels with 3·25 × 18 tyres and 6 in. brakes. At the Earls Court Show a special version with selfstarter was exhibited, the starter motor being mounted just under the tank in front of the cylinder head. It was connected by belt to the magneto flywheel and powered from batteries carried in two large boxes mounted pannier fashion either side of the rear wheel.

For 1954 there were some changes with the Supreme changing its engine for the 224 cc 1H unit and its frame for one with swinging fork rear suspension. The Embassy gained plunger rear suspension, while the Popular, Sidecar and selfstarter models continued as they were but fitted with the new 8E power unit. The Popular finally gave up its girders for 1955 when it was fitted with telescopic front forks, and the swinging fork Supreme was joined by the Envoy, a similar model fitted with the 8E engine. Both these models had larger fuel tanks fitted with a luggage grid. The Embassy continued and was made available in three- or four-speed form and the first of these continued to be offered with the selfstarter. The sidecar model continued in the range with its girders.

The 1956 lineup saw a good few changes and some streamlining of the range with the Embassy, selfstarter, sidecar and Popular with 8E engine all being discontinued. The last was replaced by a new model fitted with the 30C Villiers engine in a swinging fork frame fitted with telescopic forks and 5 in. brakes. The Envoy continued with the three-speed 8E unit and was also made available with a 9E unit with four speeds. In both cases a new tank, headlamp and dualseat were fitted.

Above **1948 Ambassador Series II fitted with the 5E twin-port engine**

Above **Ambassador Supreme with 6E engine from 1951. Note flying wing engine emblem**

Below **1950 Ambassador Series IV or Popular with 6E engine**

Above **Envoy Ambassador built from 1954–58 using 8E engine**

Above right **1959 Ambassador Popular fitted with 2L engine, earlier called Statesman**

Top **1954 Ambassador Supreme with 1H engine**

The Supreme was continued with new forks and full width light alloy hubs in both wheels. The headlamp and dualseat were also new and the rear chainguard made deeper. A Burgess silencer and Girling rear units were fitted, while the footrest position was altered to improve cornering clearance.

Towards the end of the year the Supreme with 1H single cylinder engine was modified to take the 250 cc, 2H engine and joined by another version using the 2T Villiers twin 250 cc unit. Other than the engine and a reduction in wheel size to 17 in. the twin machine remained much as before but with a performance of just over 70 mph available. Also fitted with a new power unit was the Popular but this was less obvious as the change was just from the 30C to the 31C. The two Envoy models continued.

For 1958 a new model was released, the 173 cc Statesman using the Villiers 2L engine with either three- or four-speed gearbox. This used the cycle parts of the Popular which continued and was also available with the choice of gearboxes. The Envoy models went to 17 in. wheels and the Supreme twin continued along with the 250 cc single.

By this time the company was also the sole importer of the Zundapp moped, motorcycles and scooters all of which helped to keep the wheels of the firm turning over.

During 1958 the 250 cc single was dropped as was the Envoy fitted with the 8E engine whose place was taken by the three-speed 9E. The Statesman became the Popular and the 150 model of that name went from the range. Thus the singles in the line-up comprised the 173 cc Popular with 2L engine and 197 cc Envoy with 9E motor. In both cases they were available with three- or four-speed gearboxes and followed the same format of swinging fork frame and telescopic forks. The Envoy was available with rectified or direct lighting but the Popular had direct only.

With these went a new version of the twin announced a little earlier. This was the Super S and, while it followed the general lines of the Supreme and was still powered by the 2T engine, there were a number of alterations. The gearing was raised a little and both brakes increased in size to 7 in. diameter while remaining of the full width, light alloy type. The petrol tank was changed and the rear of the machine semi-enclosed by a skirt which extended from the air filter to the rear number plate and down from the dualseat to cover the fork pivot while leaving much of the wheel in view. Similar to the Triumph bathtub in fact. Near the front on each side a horn was built in.

The petrol tank was fully rubber mounted and quickly removeable in a manner used by many machines in the 1980s. The seat was retained by two coin turned screws and under it went battery, coils, tools and tyre pump. The front forks were of a new design with two springs in each leg one of which could be pre-set for load to suit the rider.

In April 1959 the Envoy was replaced by the 3 Star Special which used the rear enclosure of the Super S on a machine fitted with the 9E engine and three or fourspeed gearbox. The styling was carried further with a massive front mudguard that was valanced in to the brake drum, a pressing over the handlebars and control cables, a grab handle behind the dualseat and three stars on each side of the rear enclosure.

The adjustable forks were fitted and 6 in. diameter full width brakes to both wheels, while a full chaincase for the rear chain was available as an option. Finish was in tartan red and black.

The model name was shortened to 3 Star for 1960 when it continued with the enclosed rear chain as standard and an optional colour scheme in greystone white and black. The Super S was also fitted with the rear chain enclosure, front mudguard and handlebar pressing from the 3 Star. Both had a folding kickstart, while crankcase covers, wheel hubs and carburettor shroud were polyfilmed. On the Super S the name, flanked by a pair of stripes was on each side of the enclosure panels.

The Popular continued unchanged to leave the range during the year but a further new model joined it for 1961. This was the Electra 75, a version of the Super S fitted with the 2T engine with Siba electric start. It had a higher compression ratio than the S and a larger carburettor. All three models had a revised tank which extended further forward to partly conceal the steering head and a revised headlamp mounting that removed the earlier side brackets from the forks.

Despite the similarities of appearance the Electra 75 was easy to distinguish as it was finished in royal gold and black to the same scheme as the others and fitted with whitewall tyres.

Also new early in 1961 was the Sports Super S which was a Super S fitted with dropped bars, flyscreen, small mudguards, abbreviated rear enclosure and a 2T engine with improved power output. In this guise it was road tested at 77 mph.

The four models continued unchanged into 1962 and were joined by a new version of the Popular fitted with a three-speed 9E unit. It was a straightforward low-priced, lightweight which did however retain some of the rear enclosure in a partial form. A deeply valanced front mudguard was fitted and many parts came from the 3 Star model, but the brakes were of 5 in. diameter and the tyre section smaller. Finish was in stone white and black with gold lining which gave the machine a smart appearance.

For 1963 the Popular was to be made available with a four-speed gearbox, while the rest of the range continued, but in October 1962 the manufacturing rights were acquired by DMW and all activity was transferred to their Dudley factory while the Ascot works were in time taken over by Page Engineering to do precision subcontract work.

With the takeover the range was contracted and the Popular dropped. The 3 Star, Electra and Super S continued with 18 in. wheels, MP hubs, 6 in. Girling brakes and DMW cowls, while the Sports Super S became the Sports Twin with similar changes but was soon dropped.

During 1963 the two ranges were revamped into one with mechanically common parts but different tank styling and badges. Thus the Ambassadors received new frames with square tubing for the main loop as was DMW practice. The 3 Star with the four-speed 9E continued as did the Super S and Electra 75. The rear enclosure remained but was made from fibreglass and the colours became black with red or white.

The three models ran on into 1964 and the two twins were fitted with the Villiers 4T engine in place of the 2T. However, DMW did not continue with the Ambassador name for long and at the end of the year the machines were dropped.

Colours

1946–49: Black with silver tank lined in red and black; chrome plated exhaust, bars and controls; black rims; stainless spokes.

1950: As 1946 but with chrome plated rims and other minor parts.

1951–52: **Supreme**—tank chrome plated with grey side panels, lined in black and red and carrying company name; cylinder barrel black; grey handlebar grips, kneegrips, footrests and brake pedal rubbers, control cables, battery and saddle cover; lighter grey frame, wheel rim centres, brakedrums, hubs, mudguards, headlamp shell, speedometer body, toolboxes, bar clips, propstand; chrome plated chainguard, saddle springs, horn, air filter, brake rod, controls, fork legs, rear sprocket, spokes and wheel rims. **Embassy** and **Courier**—black with grey tank lined red and black; chrome plated wheel rims, chainguard. **Popular**—as **Courier** except black rims lined red.

1953: **Supreme**, **Embassy** and **Popular**—as 1951. Sidecar outfit—machine as **Embassy**, sidecar black with red upholstery.

1955: **Popular**—red, silver tank. **Supreme**—grey, red or black, silver tank. **Embassy**—silver grey tank, silver rims, black for other painted items.

1956: **Envoy** and **Supreme**—deep maroon mudguards, toolbox, tank which had chrome panels, and headlamp. **Popular**—maroon all parts except tank in silver-grey.

1958: **Statesman**—black with chrome tank panels. **Popular**—all black. **Envoy**—black with gold lining. **Supreme**—as 1956 but in black.

1959: **Popular**—black. **Envoy**—black with chrome tank panels. **Super S**—red tank with black lined panels, red rear enclosure with black lined flash on each side, red forks and front mudguard; black frame; chrome plated wheel rims. **3 Star Special**—tartan red forks, headlamp shell, mudguards, rear enclosure and tank which had black, gold lined, panels; black frame; cream piping to panels; black seat; chrome plated star motifs and wheel rims.

1960: **3 Star**— as 1959 or greystone white for

Above **The Ambassador 3 Star Special with enclosure, very valanced front guard and 9E**

Above right **1962 Ambassador Popular outside Ascot works**

Top **1959 Ambassador Super S twin with 2T unit and rear enclosure**

mudguards, rear enclosure, forks, headlamp and tank; black tank panels and frame. **Popular** and **Super S**—as 1959.

1961: **3 Star** and **Super S**—as 1960. **Electra 75**—royal gold and black in style as other models; whitewall tyres. **Sports Super S**—as **Super S** in greystone white and black.

1962: All models—as 1961. **Popular**—stone white mudguards, enclosure, forks, headlamp and tank which had gold lined black panels; black frame and chainguard.

1963: **3 Star**, **Super S**, **Electra 75**—mudguards, enclosure, headlamp and tank in white or red; tank panel, frame and forks in black.

1964: All models—as 1963.

BAC

When the Bond Aircraft and Eng. Co. passed their Bond Minibyke design to Ellis of Leeds they went into production of two conventional lightweights. Both machines used a rigid welded frame with single main loop and two chainstays. The front forks were telescopic and of BAC design.

The machine was quite small but the proportions were kept in correct relationship so the appearance was normal. The wheels were shod with 2·00 × 20 tyres and had 3·5 in. diameter brakes and the machine was equipped with direct lighting, deep chainguard, Lycett saddle, and tyre pump clipped to the front down tube.

Power units were either the two speed Villiers 1F of 98 cc or the three-speed 125 cc JAP.

The two machines, both called Lilliput, failed to last for long and the 125 was dropped at the end of 1951 while the 98 cc model continued into 1952 before going. In their place came 98 and 125 cc Gazelle scooters with 1F and 10D engines and odd appearance. This arose because the fuel tank was of normal motorcycle shape but mounted over the rear wheel and the engine was enclosed by six horizontal bars spaced from floorboard to seat to make up a protective grill.

The machine failed to catch the public eye or purse and production ceased at Bond during 1953.

The design was taken over by Projects and Developments of Blackburn and the fuel tank moved forward, but no more was heard of the machine.

Colours

1951: **98 cc**—polychromatic bronze, tank lined in red with red monogram on each side. Chrome plated wheel rims, bars, exhaust system, headlamp rim and controls.

Bond

Company names and changes can often be hard to follow and in the early 1950s Bond was a name that caused a good deal of confusion. The basic facts are that Lawrence Bond was a designer and responsible for both a three-wheeler and a motorcycle, both a little unusual in appearance.

The motorcycle, when launched in 1949, was built by Bond Aircraft and Engineering Co. Ltd., of Longridge in Lancashire. At the end of 1950 manufacture was taken over by Ellis of Leeds and the original company produced a second range of machines using the same engines and sold

BAC Lilliput with 125 cc JAP engine. Whole machine built on small scale

BAC Gazelle with 10D engine in cage and rear tank changed from original

under the BAC label. At the same time the three-wheelers, called Bond, were built by Sharps Commercials of Preston. While it lasted it was confusing.

Two prototypes were first shown to the press late in 1949, one fitted with the 98 cc Villiers 1F, two speed engine unit and the other with the 125 cc JAP unit. It was the frame construction that was unusual for it was based on the use of aluminium as a riveted stressed skin. The main member was of oval shape and tapered back from headstock to rear wheel. It was formed from 18 gauge sheet and lap joined underneath using pop rivets. At the rear end it was cut away in the middle to give room for the wheel and the edges flanged to mate with the rear mudguard.

The mudguard comprised two sheet spinnings welded together to give a form that enclosed over half the wheel. It was riveted to the main member and the wheel mounting was into vertical slots in a steel strip bent in a 'U' and riveted to the lower edge of the rear mudguard.

The engine sat in a loop formed from a flat steel strip which bolted to the underside of the main beam just behind the headstock and in front of the rear mudguard. To it were welded the engine mounting lugs and two cross bars which supported the combined legshields and footboards.

Rear chain adjustment was distinctly odd and done with a jacking screw fitted between the strip riveted to the rear mudguard that carried the wheel and the main loop. By turning it the loop was pulled within its normal spring to move the engine and so adjust the chain.

The front of the main beam was closed by an aluminium casting bolted into position but not before a bracket for the saddle nose was fitted and the 1·5 gallon fuel tank located within it. The casting carried the plain bearing steering head and the forks which were formed from steel strips attached to the front mudguard which was as deep as the rear one. The relatively weak section just below the steering head led to a quick

Bond Minibyke with stressed skin aluminium frame and vast mudguards

reassessment and for production tubular forks were fitted. These were without springing, which was looked after by the use of large section tyres.

The wheels were of Bond design with split rims carrying 4·00 × 16 in. tyres and 4 in. brake drums in alloy hubs. The rear was operated by the rider's heel and the front by the normal lever. Equipment included direct lighting, bulb horn and a luggage carrier over the rear wheel.

Only the 98 cc model was offered for sale at first. In July 1950 the comments of the press when trying the prototype resulted in a change to telescopic front forks, while the very large mudguard was retained. At the end of the year the 98 cc model was joined by the 125 and both went into a modified frame loop of steel tubing in place of the strip. Both machines continued to be finished in polychromatic light blue as was standard for the make for its life. At the end other colours were offered.

The two models continued unchanged into 1953 but production of the smaller ceased in January and that of the 125 in August.

Bown

This make dates back to the 1920s but had long since vanished when it reappeared as the new name for the Aberdale autocycle.

The new name and machine were announced early in 1950 and differed in several ways from the Aberdale. First the engine was the new Villiers 2F single-speed unit, and to suit its mountings a cradle frame was designed with duplex down tubes.

The general construction followed standard autocycle lines with the petrol tank fitted between the frame tubes and its lines continued down in the form of engine shields. The front forks were blade girder and no rear suspension was fitted. A saddle, luggage grid, lighting equipment and a rear stand were fitted and the finish was in gold-lined maroon, the wheel rims being chrome plated.

A year later the firm added a pair of 98 cc motorcycles to the continued autocycle, both these being powered by the two-speed Villiers 1F unit. This was fitted into a cradle frame with duplex down tubes which extended under the engine unit. The rear was rigid and at the front the suspension was by tubular forks moving on girder links. Dunlop wheels and tyres were used and fittings included a saddle, cylindrical toolbox clipped to the seat tube, nicely proportioned fuel tank and direct lighting for the standard model, while the de luxe had the benefit of a battery. Finish was the usual Bown maroon with gold lines, and as well as the usual optional speedometer it was also possible to fit it with a pillion seat and rests.

While it represented minimal motorcycling, the Welsh machine was sturdily built and able to reach about 40 mph ridden solo. Consumption varied according to use, but figures ranging from 120 to 180 mpg were recorded so its appetite could be considered frugal.

The lower of the two gears, controlled by a hand lever on the bars, was decently high and so useful for hill climbing as it allowed the machine to nearly reach 30 mph. Braking was good and the machine stable to ride.

Encouraged by the reception of their machines, Bown moved up a class and introduced the 122 cc Tourist Trophy model in June 1952. This used the Villiers 10D engine with three-speed foot change gearbox and the unit was fitted into a full cradle frame with duplex down tubes. Like the smaller model it was rigid at the rear but had MP telescopic forks fitted at the front, these having rubber gaiters.

The tank and brakes were larger and the cylindrical toolbox mounted across the frame just under the rear of the tank rail. The machine was fitted with direct lighting, bulb horn and saddle, and was finished in maroon with blue-grey panels lined in gold.

It was a smart model and when road tested it recorded 49 mph and around 100 mpg. The brakes worked well and it proved comfortable even on long runs of a distance more associated with 500 cc than 125.

Late in 1952 the models for the next year were

Front page of Bown catalogue showing the model with 1F engine

announced and all four continued in the same form and finish. The only real change lay in the engines of the 98 cc motorcycles which were changed from the 1F to the 4F. The de luxe version had further features added in the form of a stop light, legshields, engine splashguard and a rear carrier.

In this form the range continued into 1954 but during that year production ceased.

Colours
1950: Maroon with gold lining on tank and panels; chrome plated wheel rims, bars, exhaust pipe and filler cap.
1951: **Autocycle** as 1950, 98 cc m/c maroon with gold lining, maroon rims; chrome plated headlamp rim, exhaust, bars and controls.
1952: 98 cc models as before; 122 cc **TT** in maroon with blue-grey tank panels gold lined. Maroon frame, forks, wheel rims, mudguards; chrome plated headlamp rim, exhaust, bars and controls.
1953/54 as 1952.

Butler

Chris Butler ran a company producing a wide range of fibreglass parts in the 1960s and in 1963 decided to go in for complete machines.

His first was a trials model fitted with a Villiers A series engine with Parkinson top half conversion. The frame was built from square section tubing as were the leading link forks for the most, only the rear loop being tubular. Swinging fork rear suspension was used and both ends were controlled by Girling spring and damper units. As was to be expected Butler Mouldings supplied the 2-gallon tank, front mudguard and combined rear guard, seat pan and air filter enclosure, all of which were in fibreglass. Motoloy hubs were used for both wheels and the weight was a commendable 210 lb.

The Butler in 1963 with fibreglass seatpan, tank and mudguards, A series engine

A year later three prototype scrambles machines were to be seen, two driven by 36A engines, one standard and the other with a Parkinson conversion, while the third used a single carburettor Starmaker unit. The frame was constructed from round tubing but retained the same forks as the trials model. Exhaust, tyres and fittings were to suit the different environment and tank and mudguards continued in fibreglass.

In the middle of 1964 two production machines were announced and both were trials models sold in kit form. They were identical aside from engine specification, the Tempest using the 32A unit with iron barrel, wide ratio gearbox and heavy flywheel, while the Fury had an alloy Parkinson head and barrel. Both machines used the square section tubing frame, the leading link forks and in the main the equipment of the first trials model, although the smaller scrambler fuel tank was fitted.

The machines continued to be built for a while and sold in kit form but this type of tax saving vanished overnight when VAT was introduced and with it went the kit bikes including the Butler.

Colours
1964: Frame and forks black; tank and mudguards white with option of red or blue.

Cheetah

This is the name given to machines built by Bob Gollner for competition use in small quantities. This began in the 1960s so it was inevitable that he should use the Villiers A series engine and, if wanted, a square barrel conversion.

Cycle parts were conventional but well executed as is so often the case with limited output machines, and the result functional.

Gollner did not stop at the Villiers engine but also built machines with Triumph Cub and Husqvarna units so was not too affected when the supply of engines from Wolverhampton dried up. In time he went on to use Japanese units to keep the Cheetah name going.

Commander

Futuristic styling, monoshock rear suspension and concealed controls were a few of the features of the three model Commander range shown at the 1952 Earls Court Show.

Essentially all three were the same with just the engine units being changed. The largest model had the 10D Villiers and the other two were of 98 cc, one fitted with the 1F and the other with the 2F and pedals to make an autocycle.

Bob Gollner on left with friends and Cheetah machines with A series, Cub and Husky engines

The cycle parts were based on a spine frame built from square section tubing that ran back from the steering head over the engine unit to the fork pivot and rear enclosure supports. This main section was built from four tubes welded and cross-braced from which an engine loop dropped and a seat support frame rose. The rear suspension fork was also built from square section tubing and the two legs ran up from the pivot before they went to the rear. Welded between them was a substantial spring cup to match one on the frame and in them was bolted a 3 in. diameter spring, 5 in. long which provided $4\frac{1}{2}$ in. of movement at the rear wheel.

The front suspension was also unusual being based on the short leading link principle with the link arms extended behind their pivots to work rubber band suspension members anchored on the fork leg. Pressings hid the working parts. The main part of the fork was built up in cycle form from square tubing supplemented by a top crown braced to the lower by two external tubes. These also carried the headlight and the top crown a tube to which was welded the handlebars.

The wheels varied from model to model with the 125 having 5 in. brakes and 3.00×19 in. tyres, the 98 cc machine 4 in. and 2.50×19 in., while the autocycle had smaller section tyres at 2.25×21 in., the 4 in. front brake and a back-pedalling rear one.

It was its bodwork which set the Commander apart, for at a time when a typical utility model looked the part in black the Commanders were streamlined and bright with plenty of chrome plating. The main frame was enclosed by steel panels to form a styling line along the machine with a major pressing being attached to it at the rear to form an enclosure to act as rear mudguard, carry the number plate, mount the saddle or optional dualseat to, and provide a compartment for battery, tools and tyre pump. This was closed by a lid on the right, the saddle was retained by a knob and the rear light was enormous for the time and blended into the enclosure.

The engine was enclosed by a grille that fitted to panels above and below the unit. These were bolted to the frame while the grille with chrome plated horizontal bars could be removed by detaching a dozen fast thread screws.

At the front a large sprung mudguard was blended into a shield in front of the steering head that carried a standard headlamp reflector behind a special, wedge-shaped, prismatic lens. This shield went up to join pressings that concealed all the handlebars except the grips and the inverted control levers that worked the concealed cables. Behind the steering head the

The strange Commander seen in late 1952, this one with 10D engine

pressings carried the speedometer head and behind that in the top of the frame covers went the light switch and then the recessed filler cap for the fuel tank, carried between the main frame tubes.

There were ingenious details to work the choke and tickler from an external knob and the finish bright but varied from model to model.

The machines were announced by the General Steel and Iron Co. Ltd. of Hayes, Middlesex, and were a brave attempt to build something new and exciting.

Sadly nothing further was heard of the project and the machines failed to reach the public.

Colours
1952: All models with frame covering in ivory with chrome plated motif along middle line on each side. Chrome plated engine grill, top front part of front mudguard, front fork top grill pressing, wheel rims, hubs, brake backplates, motif on top of main frame beam. Front mudguard, engine grill mounting and rear enclosure in model colours of dark blue for 122 cc **III**, maroon for 98 cc **II** and light blue for autocycle model **I**. Saddle covers and dualseat options to match.

Corgi

During the war there arose an urgent need to provide paratroops with a means of rapid movement on landing either to assemble, disperse, or move to a more advantageous combat position. Any thought of using a conventional lightweight motorcycle was ruled out by the need for the machine, ready to use, to fit into a tubular parachute-equipped container no more than 15 in. in diameter.

To solve this problem the designer Lt-Colonel J.R.V. Dolphin used 12 in. wheels, a horizontal single cylinder two-stroke engine, telescopic saddle pillar and folding handlebars and footrests. On landing the soldier pulled up the seat and bars, locked them and folded out the footrests. Within a few minutes from landing he could be a couple of miles from his drop position ready for action.

The Welbike, as the army machine was called, was very basic as it was considered to be expendable and there was no call for rider comfort. For civilian use something a little more civilised was called for and the result was the Corgi.

The engine was the single speed Excelsior Spryt unit of 98 cc which was installed in a fully duplex frame. From the headstock two tubes ran back parallel to the ground to the rear of the machine. They then curved down and round to carry the rear wheel lugs and then run forward under the engine at which point they slanted up to rejoin the headstock. There were cross braces which supported the engine and seat.

Unlike the Welbike which had a pressure fuel feed, the Corgi carried its fuel in a normal style tank above the engine and had a screw type air vent in the filler cap to prevent leakage when the machine was in transit in another vehicle. On the exhaust side the pipe ran along the right side and then curved up to a cylindrical silencer mounted across the frame under the upper tubes and just ahead of the rear wheel. From this an outlet, also on the right fed the gases into a second silencer at wheel top level.

The front forks were unsprung and above them rose tall swept-back handlebars. These were connected to both fork yokes being locked to the lower one by spring-loaded pins and passing through the upper in pivots. Thus release of the pins allowed them to fold down. The footrests simply folded up for stowage, while the saddle was mounted on a stem which could be dropped down into a vertical tube in the frame for the same purpose. When up it was clamped in place and a location pin gave the choice of two saddle heights.

The wheels were originally of normal wire

Right **Excelsior Spryt engine, built by Brockhouse and used in the Corgi**

Below **The Welbike and its drop container being unloaded while another rides (wobbles) by (IWM)**

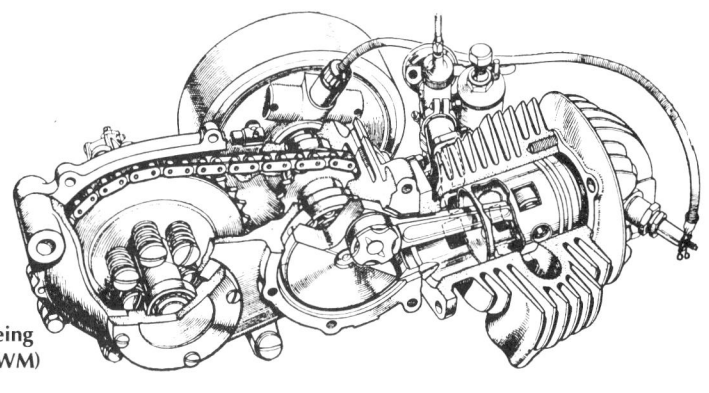

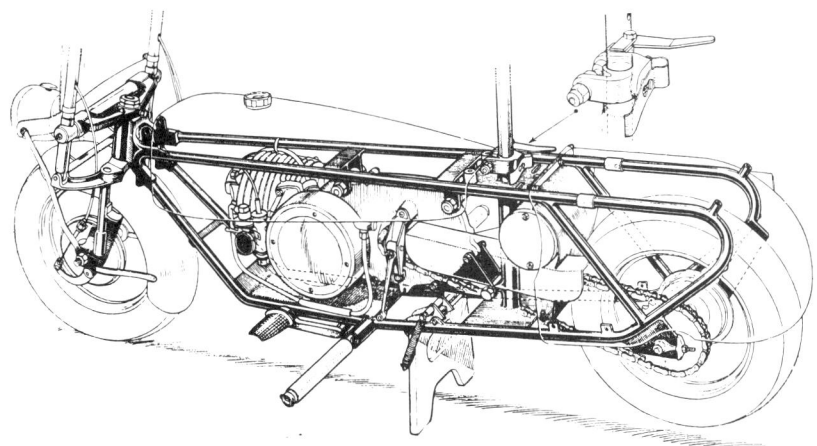

Villiers Singles & Twins

The civilian Corgi with fully duplex frame and fold-up features

45

spoke type but were soon altered to disc type and carried special 2·25 × 12·5 in. Dunlop tyres. Each had a 4 in. diameter single leading-shoe brake. The front mudguard had a deep section so gave good protection, while at the rear a sheet metal enclosure was used to cover the wheel down to spindle level. This had an internal front section to keep road dirt off the first silencer and a flat top for parcels. It was quickly detachable and under it went a chainguard on the left.

Equipment was basic but complete with centre stand, lights, bulb horn and optional shopping basket to fit to the handlebars while the detail work was good with aircraft locking nuts used at important points.

While the Corgi was first described in *Motor Cycling* in March 1946, it was 1948 before the Mark 1 became available to the public. It was made by Brockhouse Engineering of Southport, who also made the engine under licence, and sold by Jack Olding from North Audley Street in London.

In those austere days customers happily accepted the need to push start the machine, which was easy enough to do, and it proved to be quite capable of using to do the shopping and for running down to the postbox. The need for push starting went in July 1948 when a kickstarter was added on the right side of the engine. At the same time a dog clutch was added into the transmission which coupled the output sprocket to the shaft. This was controlled by hinging the right footrest up which was needed to give clearance for the kickstart. It was a useful improvement.

With these changes came a small box sidecar which the machine proved capable of pulling and was a boon for shopping expeditions and for carrying fair sized loads. The chassis attached at three points and had a sprung and enclosed sidecar wheel, while the steel body had a snap fastened canvas top.

The Corgi with dog clutch was known as the Mark II and in the middle of 1949 both models were offered a two speed gearbox and telescopic front fork conversion by Aston Bros of Coventry. The gearbox was an Albion J2 with kickstarter and its change lever was linked to a foot pedal on the right to give down for bottom and up for top. The gearbox went behind the Spryt engine with provision for primary chain adjustment and, as it occupied the space used by the silencer, this was relocated and a new rear chainguard fitted.

The telescopic fork was of simple tubular construction which gave 1·5 in. movement but it was damped using brake fluid.

A year later in 1950 Jack Olding had a scooter-type bodywork on offer for the Corgi which fully enclosed the rear of the machine up to the seat with footboards and legshields. A new fuel tank was fitted within the bodywork and a small compartment with lid was formed in the rear just above the number plate.

The machine plus its options continued into 1951 with the addition of a banking sidecar for goods carriage. This had a styled sheet-metal body which could carry up to 60 lb of goods and these could be covered by a lid which hinged at the rear with a front catch. There were no banking stops and the sidecar would fit the standard Corgi or the one with the scooter-style body.

For 1952 the Mark IV was introduced with the two speed gearbox and telescopic forks as standard. The machine was also fitted with a weather-shield clipped to the handlebar risers and this carried a larger headlamp. Other fitments in its specification were a tank top parcel grid, larger saddle and footrests, electric horn, hinged rear mudguard to aid wheel removal and a hand adjuster for the rear brake. The more utility Mark II continued alongside the new model for another year before it was dropped.

The Mark IV continued with a larger headlight during 1953 but public taste for such a machine was fading and in 1954 production ceased.

Colours
1948: black frame, forks, mudguards, headlamp shell; maroon tank; silver sheen handlebars, silencers
1953: black with gold lining, polychromatic bronze, turquoise or Indian red

Cotton

F.W. Cotton founded his firm at Gloucester after the first World war based on a frame design patented in 1914. This was fully triangulated and built using straight tubing only with the result it gave exceptional handling.

The marque sprang into prominence when Stanley Woods won his first TT on one in 1923 after a fifth the year before despite a machine fire during a pitstop. Production, using proprietary engines, reached a peak in the twenties but then declined and ceased just after the second war.

It restarted in 1954 following a company reorganisation and the first new machine was called the Vulcan and powered by an 8E engine. This went into a frame with many of the straight tube features of the past and it had four running from headstock to the rigid rear wheel and duplex loops under the engine and again back to the rear wheel. MP telescopics went at the front and the machine had rectified lighting, electric horn and a dualseat.

It was joined in 1955 by a model called the Cotanza fitted with the 242 cc Anzani engine in a frame with swinging fork rear suspension. Good sized brakes were used by both machines which were joined during the year by a Vulcan with three speed 9E unit in the twin cycle parts.

Two further models were announced late in 1955, both using the swinging fork frame, one fitted with the four speed 9E unit and the other with the 322 cc Anzani twin unit. This last had a two-into-one exhaust system, twin toolboxes and full width hubs. A trials model using the 9E engine was also added to the range for 1956.

The Vulcan with 8E was dropped during the year and for 1957 the range comprised the standard and trials models with 9E unit plus the two twins with Anzani engines. These were joined by the one new model, the Villiers Twin, fitted with the 2T engine in the Cotanza 325 cycle

Cotton Trials 197 with 9E engine from 1959–63 period

Part Two Machines

Above **Cotton Double Gloucester sports model based on Herald which shared 2T engine**

Below **Standard Cotton Herald twin of 1959–63 period with rear enclosure and leading links**

Above **Cotton Trials Special with 36A engine plus Parkinson conversion, built 1963–65**

parts. This range continued for 1958, when the standard Vulcan was fitted with Armstrong leading link forks, and into 1959 when the Villiers Twin was renamed the Herald.

As the Herald it also gained the leading link forks which were fitted to the whole range that year and rear enclosure panels which went onto the standard Vulcan as well. It was joined by a larger version, the Messenger, fitted with the 3T unit and the rear enclosure.

The Cotanza models went to special order only for 1960 and were dropped during the year, while the remainder of the range continued and was joined by a new competition mount, the Scrambler, powered by a 33A engine. The frame was similar to that used for the road models but had added bracing as did the leading link forks behind the wheel.

In March 1960 the range was augmented by the Double Gloucester, a sports version of the Herald powered by the same 2T engine in the same frame. The forks were based on those used on the Scrambler and the rear enclosure was much reduced. Equipment was sporting with dropped bars, flyscreen and narrow mudguards.

For 1961 the sports twin was joined by the Vulcan Sports using the same cycle parts with the 9E engine. The other models all continued, the standard Vulcan with the option of full or shortened rear enclosure, and were joined by the Trials 250 fitted with the 32A engine. The Scrambler became available with the option of the 34A unit and another 250 twin was introduced as the Continental. This was a Double Gloucester fitted with a tuned engine, duplex frame and bigger brakes. In April 1961 it was joined by the Corsair, a single cylinder version fitted with the 31A engine, single loop frame, larger brakes, normal bars, partial rear enclosure and a small headlamp surround-cum-flyscreen.

This range was continued for 1962 with the Scrambler fitted with the 34A engine only and joined by a cousin, the Cougar scrambler. This had a new duplex frame which housed an engine based on the 34A but fitted with a Cross barrel and piston. These were most unusual for the barrel was in aluminium and neither linered nor hard chromed. The piston ran direct in the bore supported by a very special piston ring and two steel bands rivetted to its skirt. The ring comprised two parts each of which had seven turns and were assembled onto the piston to provide multiple knife-edge contacts separated by the wider ring which acted as a ring land.

Further unusual features were thread inserts for the head bolts, a 45 degree split ring clamped down by the head to form a seal to the barrel, and a similar arrangement between the exhaust pipe adaptor and the port.

The remainder of the machine was straightforward with leading link front forks, taper roller head races and new full width alloy hubs, the rear incorporating a cush drive.

In June an experimental Cougar was seen fitted with the Starmaker engine and month later a Cotton was run in a road race using a modified 34A engine. The result in the 1963 range was two more models both using the new Starmaker unit. The scrambler was called the Cobra and used the Cougar cycle parts, while the road racer was the Telstar using similar parts suitably modified with the machine equipped with a fairing. The other models continued much as they were, although the Messenger received the duplex frame used by the Continental and a change to a 19 in. front wheel. A Continental Sports model was added with larger carburettor and new hubs plus plated mudguards.

With the new Cobra there was no reason to keep the Scrambler in the lists, while the Cougar engine was changed to the 36A with a Parkinson head and barrel. New was the 250 Trials Special **also with a Parkinson on a 32A bottom half.**

During 1963 Cotton began to sell machines in kit form commencing with the two scrambles models and for 1964 reduced their range by dropping the Vulcan standard, Vulcan trials, Herald, Double Gloucester and Continental. Of

Firing up the Cotton Conquest, a 1965–68 model using the road Starmaker engine

Conquest with optional telescopic forks and in trail machine format

the continued machines the Messenger and Corsair went in the middle of 1964 but the Vulcan and Continental Sports models carried on.

On the competition side the two trials and two scrambles machines received detail changes, while the Telstar gained the benefits of Derek Minter who was to ride the works development model in British races. His expertise made the machine into a very good club racer indeed and by June 1964 it was known as the Mark 2 and fitted with a 7 in. twin leading-shoe front brake.

The 1965 range was expanded with the addition of road and trials machines powered by the Starmaker engine so that, with the Cobra and Telstar, Cotton were able to line up four distinct models with the same basic engine. Among the other machines the engine of the Continental Sports was changed to the 4T, although the 2T remained as an option and in fact the other twins were still built for export or to special order.

The works Telstar was fitted with a six speed gearbox which helped Derek Minter if no-one else in his races, and a 360 cc scrambler was tried out.

For 1966 most of the range carried on except for the Trials Special and the Cobra which was replaced by the Cobra Special with new forks, a lighter frame and fibreglass tank and mudguards. 1966 saw the end of the Cougar, while for 1967 the 250 Trials became the Trials 37A fitted with the Villiers engine of that number in the Starmaker Trials cycle parts.

The Vulcan with 9E, Continental Sports and all four machines with Starmaker engines continued with the Telstar fitted with a new lighter frame based on the works machine and 18 in. wheels. Telescopic front forks became available as an option for any of the trials or scrambles machines.

Cotton were of course totally dependant on Villiers for engines and it was this which was to bring them to their knees in the next year or two. For 1968 they continued with their machines, the Cobra renamed the Cossack and sold in kit form, but during the year their supplies dried up and production ground to a halt.

It revived in the 1970s using foreign engines and in 1980 the name reappeared on the tank of a road racer fitted with a Rotax 250 cc tandem twin engine. In time this became successful but by then it carried another name on the tank and there were no more 'Cotton reels' to be had.

Colours
1954: **Vulcan** (8E)—black; chrome wheel rims.
1956: **Vulcan**—red and chrome. **Cotanza 325**—golden brown; lined tank; chrome plated wheel rims.
1957: All models—black, maroon or red.
1958: All models—maroon or black to special order.
1962: **Cougar**—pale green. **Corsair**—light blue; black tank panels, frame and headlamp shell. **Vulcan Sports**—red and black; chrome plated mudguards.
1963: **250 Trials Special**—frame black; tank red with black panels.
1964: All models: red and black.

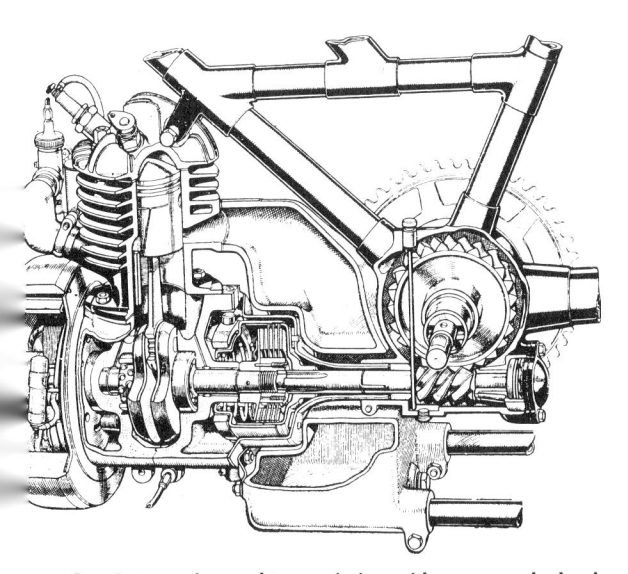

Cyc-Auto engine and transmission with worm and wheel reduction gear

Cyc-Auto

The design of this machine, an autocycle, goes back to 1931 and from the start it was a little different to most as, although the engine was mounted with the cylinder vertical, the crankshaft axis lay along the machine. It drove into the bottom bracket via a worm and wheel from where a chain took the drive to the rear wheel.

At first Cyc-Auto engines were fitted but for 1938 a change was made to Villiers units. The company then ran into financial problems and was sold to the Scott concern, who redesigned the power unit and moved the firm to premises in Acton, West London. The engines were made at the main Scott works at Shipley in Yorkshire and at first the Villiers unit remained as an option and in 1939 both ladies and gents models were available.

After the war a single machine with the Shipley-built engine was the only one to continue in production. The engine was based on the same principles as the earliest unit and had dimensions of 50 × 50 mm and a 5·3:1 compression ratio with crankshaft laid fore and aft. This was of pressed up construction with bobweights and ran on ballraces front and rear with close fitting bushes fed with oil providing the necessary gas seals. The crankcase was split vertically and the iron barrel with integral head bolted to it. At the front went the flywheel magneto to provide lights and ignition and at the rear a multi-plate clutch whose withdrawal applied a transmission brake.

The clutch drove a shaft which was coupled to the steel worm which ran in bearings in the base of the large diameter bottom bracket housing. This contained the bronze worm wheel and through the centre of this ran the pedal shaft. The gearbox had its own oil level dipstick and drain plug. Surrounding the clutch housing and drive shaft to the bottom bracket was a large cast aluminium box which was connected to the

Cyc-Auto model from 1934, hence magneto on front of crankcase

exhaust port and acted as a silencer. Twin tailpipes emerged from its rear lower corners to carry the gases away. The Amal carburettor was clipped to a right angle inlet stub itself clipped to the inlet port at the front of the cylinder.

The rest of the machine followed autocycle practice with an open frame to which the engine unit was bolted. Pressed steel front forks were fitted and fuel was carried in a cylindrical tank mounted behind the saddle.

In 1947 a detachable alloy cylinder head was adopted and the number of transfer ports reduced from three to two as by doing this the two could be more accurately made. The head was fitted with a decompressor while the inlet stub was taken as low as the flywheel allowed and the inlet manifold was made to have an internal size larger than carburettor or inlet port. This was stated to improve mixing and reduce blowback.

At the beginning of 1949 a commercial version, the Carrier, joined the standard model and was fitted with strutted bicycle forks and a large basket carrier over the front wheel.

For 1950 there were some changes to both engine and cycle parts. The Autocycle was replaced by the Superior model which had a fuel tank mounted between the frame tubes in the usual autocycle manner but otherwise continued in much the same form as before, as did the Carrier model.

Both had the engine revised so that the carburettor was mounted at the rear of the cylinder and twin exhaust pipes curled down from the front to run along under the engine and bottom bracket to small silencers on each side of the rear wheel.

For 1952 the Carrier was fitted with the pressed steel forks of the Superior but otherwise the two machines continued unchanged to the late 1950s when they were discontinued as the market had turned to imported mopeds.

Colours
1946/49: generally black.
1950: **Superior**: silver grey mudguards, chainguard and tank which was lined in red; black frame, forks and silencers; chrome plated exhaust pipes, wheel rims, bars and controls.
Carrier: black, chrome as **Superior**.
1951 on: **Superior**: chrome plated silencers.
Carrier: as 1950.

DMW

Dawson's Motors, Wolverhampton, produced the initials first seen on a couple of grass bikes just before the war in 1938. During it the owner did some work on suspension systems, and after it in 1947 a prototype Calthorpe revived that old name but was built at DMW Motorcycles. That machine used a 5E engine in a rigid frame with simple telescopic forks and a bicycle saddle.

Late in 1950 the true DMW story began with the announcement of a small range of lightweights using Villiers 10D or 6E engines. What was unusual was that the de luxe versions had frames built from square section tubing. They also had plunger rear suspension and Metal Profile front forks and were accompanied by standard rigid frame models built using round tubing.

The standard models continued with the saddle mounted on a stem, bicycle fashion, but the de luxe featured a dualseat, the front part of which doubled as the lid for the toolbox. Wing nuts at the rear allowed it to be removed for access and the de luxe had a battery, rectified lighting and electric horn. Also listed but not ever seen was a model fitted with the Villiers 1F, two-speed unit.

1952 brought no changes to the road models but they were joined by competition versions available with either engine and rigid or plunger rear ends. No lights were fitted although they were available as an extra, and trials tyres went on the wheels.

By that time the works were already running a machine of their own in trials including the Scottish and this had a four-speed box and air forks. The production models carried on with little change. The de luxe machines gained cowls for their headlamps and these carried speedometer, ammeter and light switch, while the rear number plate became boxed in. For competition, only spring frame versions were listed although the rigid ones were still available to order.

At the 1952 Earls Court Show the standard 125 was replaced by the Coronation to celebrate the event due in May 1953 and, while the machine was essentially as before with rigid round tube frame, the front forks were a new form from Metal Profiles and called bottom link. Each fork leg was a tube with pivot at its base on which went a short forward arm which carried the wheel spindle. Between pivot and wheel went a spring unit anchored to the fork leg and on the right a second link acted as a brake torque arm.

It was only built that year and at the end of it all the 125 cc models were dropped. The range for 1954 brought considerable changes with a new frame and new engines. The 6E was replaced by the 8E in the 200 de luxe, which continued with its plunger frame, while the 4S competition model was fitted with the 7E with three- or four-speed gearbox.

The new frame had a number of unusual features and was built up from a square section tubing and steel pressings. The tubing was used for the single top and down rails with the latter spreading into a wider section beneath the engine. It and the top tube joined to a set of pressings which formed the rear engine plates, fork pivot, rear mudguard, enclosure for battery and tools plus the mounting for the dualseat.

The rear fork was also unusual with oval section arms and a snail cam mechanism near the pivot for rear chain adjustment, At the front went MP forks.

Into this frame went a choice of 8E or 1H road engines, or the 7E competition unit for the Moto-Cross model. Equipment was to suit the purpose with a three-gallon tank and tapered valance front mudguard on the road models, and sprung guard, braced bars, special footrests, different seat and sports tyres on the competition one. The model with the 1H engine was called the Cortina and the same frame was also used for two models fitted with the French AMC engines.

For 1955 most of the range continued unchanged but the scrambles model was fitted with

Part Two Machines

Above **DMW 200 de luxe in 1952 with plunger frame and under-seat toolbox**

Above **DMW Mk 8 from 1957–58, a cut price model using the old 8E engine**

Below **Bernal Osborne testing the 1957 DMW Dolomite II with 2T engine**

Earles forks and became the Moto-Cross Mk 5 and the trials model became the 5S. The plunger and swinging fork 200s were joined by a pair of 150s using the 29C engine with its four-speed gearbox and the model was given the name Leda. The road models, including the Cortina, had a modified headlamp cowl which carried the horn in its right side. A further model introduced was a racing 125 using the AMC twin camshaft single and called the Hornet. All Models were available fitted with Earles forks if desired.

During the year the two plunger framed models were dropped and for 1956 the 150P continued along with the Cortina, 5S and M-X Mk 5, while the 200P became the 200P Mk 1. They were joined by a trio of models fitted with the 9E engine and equipped in the style of the older models for road use as the 200P Mk 9, for scrambles as the M-X Mk 6 and as the Mk 7 trials. Otherwise there were only detail changes.

For 1957 there were new models at each end of the price scale with the Mk 8 a cut-price model with tubular frame, three-speed 8E engine and conventional construction at one end, and the Dolomite II at the other. The latter used the 2T twin engine in the P-type frame common to much of the range. Also new with the same frame were the 150P Mk 9 and the 175P Mk 9. The first used the 31C and was in the range for a few months, while the second lasted a little longer to the middle of 1957 and was 2L propelled.

Out from the range went the Leda, M-X Mk 5 and the 5S, while the others continued but only until the autumn of 1957 when the 200P Mk 1, M-X Mk 6, Mk 7 trials and Cortina were all discontinued, the last as it had been superseded by the Dolomite.

This left the 8E and 9E models of 200 cc plus the Dolomite which was joined by a competition version with high level pipes. This was the Mk 10 and fitted with Earles forks and suitable tyres and equipment for trials riding. The Mk 8 was withdrawn during 1958 so only three models were available for 1959.

Late in that year were joined by three more, two of them for competition. The third was a larger version of the twin using the 3T engine and called the Dolomite IIA. The others were the Mk 12 trials and scrambles competition models both of which used the A series engine, the first being fitted with the 32A unit and the second with the 33A, and in either case the 9E was an alternative option.

Within a few weeks all models had a change to the wheel hubs and a new type of DMW brake using the S cam technique from trucks for operation and a wedge from Girling for adjustment. The diameter remained at 6 in. with $1\frac{1}{4}$ in. wide shoes and the brake cam was formed so that it moved out two pistons when it was rotated and it was these that forced the shoes into contact with the drum. At the pivot end a conical adjustor with flats was screwed into the backplate to force two tappets and hence the shoes outward for adjustment. The rear hub was qd and the front had a knockout spindle.

This range of six models were called K frame machines and continued through 1960 and well into the next year when the Mk 10 and both Mk 12 models were dropped. The three road machines continued unchanged and were joined by two new competition models with new frames and forks. The frames were based on a large diameter top tube, tapered down tube and an H section pressing acting as a seat member and stone guard.

The engine sat in a pair of plates which joined the frame together and linked to the rear fork pivot. The fork was pivoted in the usual manner and chain adjustment carried out by moving the rear wheel. The scrambles machine had a cush-drive rear hub with brake drum on the right while the trials iron kept it on the left. Straight tube Earles forks were fitted. Engines were the 32A for the Mk 15 trials model and the 34A for the scrambler. Also available was a DMW square barrel conversion with special cylinder head and alloy barrel fitted with liner. Earles forks continued to be available for all models although the roadsters were supplied with telescopics as standard.

During 1962 the 324 cc Dolomite IIA was discontinued along with the Mk 14 and Mk 15 competition models while the Dolomite II was joined by a Sports Twin version with higher compression engine, alloy mudguards and dropped bars. Before this happened both competition models were made available in kit form which enabled them to be offered at a reduced price assisted by the purchase tax saving.

Late in the year came the announcement that DMW had taken over Ambassador and this was followed by the introduction of the M series of machines comprising 200P Mk 9, Dolomite and Sports Twin. Each of these was much as their predecessor but with detail changes. With them were introduced three competition machines, the Mk 17 trials model with 32A engine plus the Mk 16 moto-cross and Hornet road racer with the Starmaker unit. There was an alloy barrel available for the Mk 17.

In the middle of 1963 a modified version of the Hornet was unveiled with an all-welded frame and new fork which retained the single central spring and damper unit anchored to the top fork crown and a stout fork leg bridge. Twin front brakes were fitted along with a number of detail changes which included an expansion chamber that lay directly beneath the engine unit and terminated in twin tail pipes, one on each side.

At the end of 1963 the Mk 16 scrambler was replaced by the Mk 18 with new duplex frame, and the standard P model 200 Mk 9, Dolomite II and Sports Twin went although the M series continued. The Mk 14 scrambler reappeared fitted with the 36A engine and having the option of an alloy barrel available. The Mk 17 and Hornet continued as did the M series twins which had an engine change to the 4T unit.

For 1965 the old type K frame reappeared fitted with the 4T engine as the Dolomite II while the standard model with the M frame was dropped. The Sports Twin version continued as did the Mk 18, Mk 17 trials and Hornet. The 200 cc model was finally dropped along with the Mk 14 and in place of the latter came the Mk 19 fitted with a 36A engine in a duplex frame and the 19D with the DMW barrel conversion.

Also in the news that year was a factory project to build a 500 cc twin road racer using coupled engines with Starmaker barrels and Alpha full flywheels. A five-speed Albion gearbox was used and the frame followed conventional lines with Hornet style front forks. A later version used Royal Enfield GP5 engines as a basis. There was talk of plans to build a production batch but the company was none too healthy and during 1965 the Mk 17, Mk 18, Mk 19 and the K framed twin were all discontinued to just leave the Sports Twin and the Hornet which became available in Mk 2 form with five-speed gearbox.

The twin went in 1966 and the Hornet the following year to bring the DMW story to a close, although the Metal Profile forks which had long been associated with them commercially continued in production.

Colours
1951: All models—all painted parts except black headlamp shell in turquoise blue, tank gold lined; painted wheel rims.
1952: Road Models—as 1951. Competition; light blue as roadsters.
1953: All models—aluminium oxide frame; blue for other items, tank gold lined; chrome plated wheel rims.
1954: All models—options in blue, black or Paris grey with chrome plated rims and new tank transfer design.
1955: Competition m/c—blue. Road m/c—Paris grey.
1956: Road m/c—blue, Paris grey or black: tanks lined in red and gold. Competition—blue. Seats—maroon with grey or black finish, beige with blue finish.
1957: As 1956 plus optional maroon and old gold

Above 1962 DMW Mk 15T with 32A engine. Note frame construction of machine behind

Above The DMW 500 cc road racing twin with Starmaker barrels later changed to Enfield GP5

Below 1963 DMW Mk 16 moto-cross model with Starmaker engine and flat exhaust

DMW Sports Twin, 1965 model with the 4T engine

for road models except **Mk 8**. **Mk 8**—Robin Hood green.
1958: **Dolomite II** and **200P Mk 9**—Paris grey or black. **Mk 8**—green. **Mk 10**—blue.
1960: Road m/c—Paris grey, blue or black; optional maroon and old gold. Competition m/c—turquoise blue.
1961/63: As 1960.

Dot

The name came from the initials of the slogan 'devoid of trouble', and the company was founded in 1903 by Harry Reed. For years he essentially was the company and rode its machines in many races. He won the multi-cylinder class of the TT in 1908 and years later was second in the 1924 sidecar event at the age of 49. The former year was good for the small firm with a second in the Junior as well plus a third in the Lightweight. They managed another couple of places in the late 1920s but, as the economic recession took hold, they were one of the many casualties and manufacture stopped around 1932.

It restarted in 1949 after a decade of building tradesmen's three wheelers, fitted with Villiers engines, with the announcement from the works at Arundel Street, Manchester, of a new machine. This was to have a 197 cc Villiers engine in a loop frame with Webb girder forks and be followed by a similar 122 cc model. In fact, although this was listed in a buyer's guide in March 1949, only a 197 cc prototype was built and only that capacity went into production.

The first machines were road models fully equipped with rectified lights, battery and electric horn, saddle, toolbox on the right attached to the rear mudguard stays, a lifting handle formed as one of the stays and a centre stand. Over the years the company was to turn to competition models and produce few for use on the road, but for 1950 the 197 was offered in two forms only with direct or rectified lighting as models 200/DS and 200/RS.

For 1951 these became the 200/DST and 200/RST fitted with telescopic front forks and were joined by a competition model available to suit trials or scrambles and called the Scrambler. The frame remained rigid with a reduced wheelbase and a two stage exhaust was fitted. A cast aluminium expansion box was fitted close to the cylinder and for trials connected to a pipe that ran along the top of the chaincase to a silencer and tailpipe. For scrambles this was removed and the box end rotated so that the pipe could point down to give an open exhaust. The standard tyres were trials but competition ones were available as an option. An easily detached headlight was supplied.

During 1951 Dot introduced a model fitted with the 250 cc Brockhouse side valve engine but this was only produced up to 1953. Possibly a more meritorious effort by the company was the entry of a team of machines in the newly introduced 125 cc TT of 1951. These used 122 cc Villiers engines laid forward so the cylinder was nearly horizontal in a swinging fork frame built from square section tubing. Although outclassed

Above **Youthful John Surtees on the 1956 Dot Mancunian with 9E engine**

Below **1952 Dot TD with 6E engine, a trials model with lights**

by the leading Italian four-strokes, they did all finish to win the Team prize for the firm. They tried again in 1952 but with less success and only one finisher.

When the 1952 programme was announced it continued the two road models but listed five separate competition ones, three for scrambles and two for trials. This was to be the company policy from then on and most years saw a range of machines offered all based on an engine and frame built up in various forms by ringing the changes on the equipment.

Thus for 1952 the scramblers were listed as the S with road equipment but no lights, the SC in stripped form and the SD with direct lighting as well as full road equipment. In a similar manner the trials model T was without lights and the TD fitted with direct lighting.

For 1953 Dot introduced a new frame with swinging fork rear suspension controlled by laid forward Girling units. The frame was used for both road competition models and increased the range of variations available. The two rigid road models continued with direct or rectified lighting but only the latter was offered with the new frame. The two rigid trials machines also continued to be listed and were joined by one with the spring frame and direct lighting. All three scramblers were only offered in the spring frame.

By the end of the year the firm had dropped the road models to concentrate on the competition scene. At the same time the engines were changed to the 8E which allowed three or four speed gearboxes to be used as required. Only the two trials machines continued as in the past with rigid frames, all the other models having the swinging fork. In addition an Earles front fork was available as an option for the whole range.

The rigid framed trials models were dropped during 1954 but the TH and THD with direct lighting continued with the THX and TDHX special trials models with 21 in. front wheels. Three scrambles models were offered for 1955

Villiers Singles & Twins

59

1966 Dot Demon Mk 3, Marcelle conversion on 34A, square frame tubes

Dot Trials Marshall of 1961 available with 197 or 246 cc engine

with the same options as in the past and were typed the SH without lights, SDH with direct lighting, and SCH stripped for action. The Earles forks continued to be offered as an option for all.

Late in 1955 Dot brought out a road machine once more, called the Mancunian, and powered by a four-speed 9E engine. Frame construction was conventional with swinging fork but at the front went leading links with short arms and long Armstrong hydraulically damped units. Girlings looked after the rear wheel. There was some attempt to provide an enclosed space under the dualseat but this did not cover any of the frame members.

For 1956 the TH and TDH trials models were dropped but the other two and the three scrambles machines carried on until late in the year when a further engine change occurred with the fitting of the 9E unit. This went into the five competition models which were joined by the works replica trials special with footrests further back, modified exhaust system, 18 in. rear tyre and altered gearing. All machines were fitted with the leading link suspension of the Mancunian as standard. There was also a suggestion of models fitted with 150 or 175 cc engines but, as with an earlier idea of using the 1H unit, nothing further was heard of it.

1958 saw a three-speed version of the Mancunian added to the lists but at the end of that year the road models were dropped and a number of larger machines added. The six competition 197s continued.

The new models comprised trials and scrambles machines of 246 cc and a trio of twins. The first used Villiers A series engines in the existing frames to give SH and SDH scrambles models, the second with direct lighting, and THX plus TDHX trials irons. Engines aside, they were replicas of the smaller machines and fitted with 31A units or the 9E with the Vale Onslow conversion.

The first of the twins was a competition model fitted with a tuned 2T engine with short length open pipes for scrambling. The frame was similar to the existing ones but altered to suit the twin engine and it carried a slim fuel tank of less capacity than usual. Front suspension continued to be by leading link forks and 6 in. brakes were fitted to both wheels, these being common to the range.

The two other twins were both powered by the 349 cc RCA engine, one being a scrambler based on the 250 twin cycle parts apart from a fatter rear tyre and the other a road version known as the Sportsman's Roadster.

Dot Demon International of 1963 with Starmaker engine and leading link forks

1966 Dot model WR with 32A engine

Nearly all this range ran on into 1960 with only the scrambles twin being dropped and minor changes being made to some of the other models. At the end of the year all the 197 cc machines went along with most of the 246 cc scramblers and the 349 cc RCA engined twins.

The machines that continued were the 246 cc scrambler SCH powered by the 34A engine and the works replica WR with the 32A. These were joined by two Trials Marshall machines of 197 and 246 cc capacity continuing the line of competition models. They were both dropped at the beginning of 1962 which brought the long line of 197 cc Dots to an end, but the 246 cc models continued.

In March 1962 the SCH was replaced by the Demon scrambler fitted with a 34A engine with a Marcelle alloy barrel and four speed close ratio gearbox in a square tube section frame with the leading link front forks. The standard engine with iron barrel was also available. The remainder of the machine followed the accepted practice of the times and the Works Replica trials model continued with it.

Only having two models did not inhibit the company, which continued into 1963 with a change to the 36A engine for the Demon scrambler. They were joined by an alloy works replica during the year which was a lightened version of the normal machine in a very workmanlike set of cycle parts. They may not have looked too stylish but the smooth underside and large mudguard clearance were certainly functional. Unlike the other models the wheels were built onto Italian full width hubs and in this way the front brake diameter was increased.

At the same time the company announced the Demon International scrambler with a new frame carrying the Starmaker engine. The main loop tube cross-section continued to be square and leading link forks were fitted along with many other items common to the scrambles machines. In practice manufacture was held in abeyance while other work was done and the model failed to reach the showrooms.

Thus for 1964 there were just two trials models and the Demon scrambler, all of which became available in kit form as well as fully assembled. For 1965 these three were joined by the Special Demon scrambler which was powered by an engine using the Dot alloy head and barrel on an Alpha crankcase. The remainder of the machine followed the lines of the standard Demon model.

1966 saw the end of the Demon but the special version became the Demon Scrambler

Villiers Singles & Twins

61

Mk 3 with a new exhaust and other modifications. In July it became the Mk 4 with changes to the gearbox, which gained a barrel cam to move the selectors, and a duplex primary chain.

For the 1967 season the scrambler was renamed the MK 5 and joined by a larger version fitted with a 360 cc engine of 78·7 × 74 mm dimensions. At the same time the Works Replica became the 250 Standard Trials and was fitted with the 37A engine.

All four models continued on into 1968 and during the year assembly ceased and the machines were only available in kit form. Engines went into short supply with the turmoil in the NVT group and so the company turned to other sources. Among these were Husqvarna and, in the 1970s, the 170 cc Minarelli engine which went into another kit bike and later a complete machine until production ceased in 1977.

Colours
1949: Prototype—black frame, forks and toolbox; cream mudguards; chrome plated tank, silencer and chainguard.
1950: As 1949 model.
1951: 197 cc models—cerise forks and mudguards; black frame; chrome as 1949. Scrambler—black frame; chrome as 1949; alloy mudguards.
1952: Road—cerise and silver. Competition—black and silver.
1953: Road—tank cerise with silver panels, cerise mudguards and frame; silver forks. Competition—black frame; chrome plated tank; alloy mudguards.
1954/55: Competition—as 1953
1956: Competition—as 1953. **Mancunian**—British Racing green or Continental red.
1957: Competition—green or black frame and forks; chrome plated tank, mudguards, chainguard, wheel rims.
1959: **250 twin**—red, black and chrome finish. **200 scrambler**—as **250 twin**.
1962: **Demon**—red, black and chrome.

Elstar

This company was best associated with speedway and grass track machines supplied in kit form to take the single cylinder dope engines made by JAP. They also made telescopic forks for grass machines using rubber band suspension and a Girling damper.

In the late 1960s a trials Elstar with Villiers engine made an appearance and was built up in the usual manner of kit machines at that time. Forks were conventional telescopic without gaiters, Girlings controlled the rear end, a Villiers engine with conversion kit propelled it, and a nicely styled fuel tank sat in front of the abbreviated dualseat.

A neat and functional machine of its time.

Excelsior

The Excelsior firm was founded in 1874 to build pennyfarthing bicycles and became involved in motorcycles in 1896. From then on they followed the vicissitudes of the industry and managed to

The Elstar with Villiers engine plus conversion in neat workmanlike package

win the 250 cc TT twice, the second time in 1933 with the four valve 'Mechanical Marvel'.

During the war the firm built the Welbike which afterwards became the Corgi and themselves had a two model lineup for 1946, on this occasion using Villiers power units as they had done in pre-war days. The two machines were the Autobyk with the Junior de Luxe engine and the Universal with 9D engine in rigid frame with girder forks. Perhaps the most odd feature of the latter machine was the gearchange lever which worked in a gate in the centre of the fuel tank. In other respects it was a conventional utility lightweight but one of the first postwar models.

In December 1946 the Autobyk was joined by a two-speed version fitted with the Goblin engine and both had neat tubular girders with rubber band suspension and the shielded engine.

For 1948 a third Autobyk, the S1 with the Spryt single speed engine, joined the other two having first become available in the previous May. In other respects the range continued unaltered.

Two new models joined the Excelsior line up for 1949, both given the name Minor and fitted with 98 or 123 cc Goblin engine units. The machines used the same cycle parts with simple rigid frame and girder forks but their appearance was completely different to that of the Universal. This was because the Goblin engines had well inclined cylinders, the seat tube was vertical while the fuel tank was hung below the top frame tube. A further change was that a cylindrical toolbox was mounted in front of the seat tube just below the tank, while the older 125 had its rectangular box behind the seat tube low down aft of the gearbox.

The Universal model was altered to use the 10D engine with foot-change in 1949 so lost the gear lever in tank slot arrangement. It was available in two forms, the U1 with direct lighting and the U2 with rectifier and battery. Both machines continued with rigid frames but were fitted with telescopic front forks. They were joined by two similar models, the Roadmaster R1 and R2 with the same electrical variations, and these were fitted with the 197 cc 6E unit. The three Autobyks continued as they were.

There were quite a number of changes for 1950 with the introduction of the new Talisman Twin, the highlight. This 243 cc machine had its engine and gearbox unit mounted in a new frame with plunger rear suspension. The latter was slightly unusual in that the wheel supports were welded to the spring boxes which slid up and down on the central rods held in the frame lugs. Each spring box contained two concentric main springs and a single shorter rebound one. The frame itself was of normal single loop type.

The twin was fitted with a 2·75 gallon fuel tank and a two-into-one exhaust system with the silencer on the left. Above it went a large toolbox matched by another on the right, while the seating was by a saddle. A pancake air filter was fitted to the carburettor and behind it, on the right, went the battery with the horn beside it on the left.

The new frame with plunger rear suspension was also fitted to the 122 and 197 cc models as standard, both of which continued with direct or rectified lighting systems, but the two Minor models were dropped. On the autocycle front the machines with Spryt and Goblin engines carried on but that with the Villiers was discontinued.

There were no changes for 1951 and few for the following year although the twin did go to a duplex primary chain and roller main bearing behind the magneto. The tyre sizes of the U and R model singles were altered a little. The twin was joined by a Sports version which had twin carburettors and no air cleaner for enhanced power but retained the two-into-one exhaust system. The gearing was lowered and it was fitted with a curious dualseat which had side pads below the rider's section to act as kneegrips for the passenger. In other respects it was as the standard model except for the finish which was beige.

Above **1952 Excelsior Universal U2 with battery and rectified lighting**

Top **Excelsior Autobyke with Villiers Junior De Luxe engine from 1946–47**

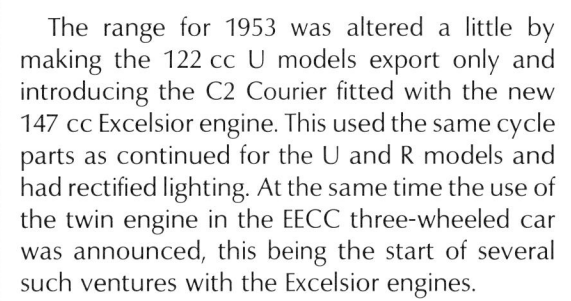

Above **Excelsior F4 Consort from 1953 with 4F and later 6F engine**

Top **1953 Excelsior Courier with their own 148 cc engine**

The range for 1953 was altered a little by making the 122 cc U models export only and introducing the C2 Courier fitted with the new 147 cc Excelsior engine. This used the same cycle parts as continued for the U and R models and had rectified lighting. At the same time the use of the twin engine in the EECC three-wheeled car was announced, this being the start of several such ventures with the Excelsior engines.

During 1953 a new model was introduced that was to continue in basic form to the last days of the company. This was the 98 cc Consort first seen as the F4 model fitted with a Villiers 4F, two-speed engine unit. It was a simple cheap lightweight with the engine installed in a rigid frame with tubular girder forks. A saddle, cylindrical toolbox, bulb horn and direct lighting completed a basic machine built for local runs.

At the end of the year the 1954 range introduced a transition period and several changes. For a while the firm listed 14 models without counting the export of only U1 and U2 which continued. Unchanged were the faithful S1 and G2, the new F4 or the C2 which was joined by a direct lighting version, the C1. The TT1 and STT1 twins also continued as did the R1 and R2, but these did have their engines changed to the 8E unit.

New were several models in a frame with swinging fork rear suspension controlled by Girling units. Two were the R3 and R4 with 8E engines and the option of three- or four-speed gearboxes and direct or rectified lighting. A second pair were the TT2 and STT2 twins and both sizes of machines had dualseats, without odd kneegrips, valanced mudguards and, on the twins, the single silencer on the right. The final new model was the Condor which comprised a Consort frame fitted with three-speed 12D or 13D engine, the choice depending on which magazine you read.

There was considerable reduction in the size of the range at the end of the year when the 1955 models were announced with ten of the existing ones leaving the lists. These were the U, R, C and twin models with plunger frames, the direct lighting R3 and the D12. This left the two Autobyks, the Consort, the 197 with rectified lighting and the two twins in the swinging fork frame.

They were joined by five further models in three sizes. The first was the Courier C3 which retained the 147 cc Excelsior engine in a new single loop swinging fork frame. This differed from the earlier type as a single tube ran round the engine unit in place of the twin under-engine rails and single down, top and seat tube retained for the R4 and the twins. A dualseat was fitted as standard along with rectified lighting.

With the C3 came the R5 and R6 Roadmaster models of 197 cc with direct or rectified lighting systems. Both were powered by the 8E engine fitted in the single loop frame used by the Courier and this frame was also used by a new version of the standard twin, the TT3. These new models had battery and toolboxes under the seat to clean up the appearance in that area. The older frame continued for the SE–STT2, a special equipment model with full width hubs incorporating 6 in. brakes front and rear.

There were further range alterations for 1956 with the R4, R5, TT2, STT2 and SE–STT2 all going, to be replaced by updated models. In most cases new colours were listed for the continuing and new machines although black remained as an option in all cases.

The two Autobyks were unchanged as was the Consort which was joined by a version with plunger rear suspension in the Excelsior style of the past. This design was also used by one of the new models, the C1 Condex which was powered by a three-speed Villiers 30C engine unit. In effect it was a replacement for the D12 with fairly basic specification but its telescopic forks were air assisted thanks to special seals which trapped air within the legs.

Lovely 1957 period picture taken at Davies & Sons in Llanidloes. Excelsior Skutabyke is the prize

Excelsior Talisman Twin Sports model from 1957, the STT5

1960 Excelsior U10 using the Villiers 31C engine not their own

The Excelsior powered C3 Courier continued in its swinging fork frame as did the R6 with the 8E engine. For the 200 cc class a new model, the A9 Autocrat was introduced which used the R6 cycle parts but the four-speed 9E engine unit. It also had a longer headlamp shell with speedometer, ammeter and light switch, while the R6 retained the fork top mounting for these items. The older model also had its exhaust pipe sweep altered, a new front stand and malleable cast hubs.

The Talisman twins continued as the standard model TT3 and sports STT4, the latter being fitted with two separate exhaust pipes and silencers, while continuing with the twin carburettors. Its instruments and light switch were mounted in a fork top panel.

In April 1956 the C3 was joined and later replaced by the C4 Convoy of very similar specification. It was lighter with smaller section tyres but retained the 147 cc Excelsior engine and three speed gearbox in a single loop frame with swinging fork rear suspension and telescopic front forks. It was fitted with rectified lighting, battery and dualseat and sold at a very competitive price.

At the same time the springer Consort model had its engine changed to the 6F with foot gearchange, while the rigid F4 continued with its Villiers 4F. This arrangement was not altered when the 1957 range was announced and the two 98 cc models were joined by a third with extensive enclosure panels and legshields combined with footboards that ran from the shields back to the rear plunger units. Unlike the Consorts the new Skutabyk, was fitted with telescopic front forks. To further improve rider protection deeply valanced mudguards were fitted over both wheels.

With three 98 cc motorcycles in their range and the many imported mopeds that had reached the market the need for autocycles had gone and so the two Autobyks were withdrawn after their long run. With them went the F4S Consort, C1 Condex and C3 Courier with their Excelsior engines, and the STT4 Sports Twin which was replaced by the STT5.

This and the standard TT3 twin were fitted with modified forks without gaiter seal but with a three-rate spring. Chainguard, kneegrips and dualseat were all improved by detail changes, as were the silencers, and the Sports model was fitted with a full chaincase and an engine with revised heads and barrels with greater fin area. The TT3 had separate exhaust systems as used on the Sports model.

Of the singles, the C4 Convoy continued but the R6 and A9 became export or to special order only before they were all dropped during 1957. Around the same time the F4 Consort and both twins went, the latter replaced by the next series as the TT4 and STT6. For 1958 these continued along with the F6S and SB1 and were joined by four further machines.

The smallest was a further 98 cc model, the CA8 Consort with 6F engine in swinging fork frame with telescopic forks and a dualseat. Next in size were two Universal models, the U8 and U8R with direct or rectified lighting, powered by Villiers 30C engines with three-speed gearboxes, telescopic forks and swinging fork rear suspension.

The final new machine was the S8, Super Talisman Twin which used the new 328 cc Excelsior engine in the duplex tubular frame of the STT6, this being more suitable than the loop frame used by the standard 250. In other respects the larger twin copied the TT4 but did have a larger section rear tyre.

1959 brought more changes for Excelsior had developed a habit of altering model numbers year by year and so the CA8 became the CA9, and the U8 pair the U9 and U9R. The F6S went as did the STT6, while the Skutabyk, standard Talisman Twin and the 328 twin continued. The last of these was joined by a Special version, the S9, which had new forks, a 7 in. front brake, 18 in. wheels and rear wheel enclosure down to spindle level. This model was further distinguished by a headlamp nacelle which carried the instruments and light switch plus a toolbox cavity set in the top of the petrol tank whose lid doubled as the filler cap.

At the bottom end of the scale the F4F Consort was added on the lines of the old F4 but with telescopic forks and the 6F engine in a rigid frame fitted with a saddle. This basic version of the CA9 was to run on with the sprung model for several years and various type numbers, and so for 1960 the two became the F10 and C10. At the same time the Skutabyk was dropped along with the U9 pair which were replaced by a U10 Universal fitted with a Villiers 31C engine unit with three-speed gearbox. In other respects it was a continuation of the Universal series but only available with rectified lighting. It was joined by a similar 197 cc model which revived the Roadmaster name as the R10 with 9E engine and four-speed gearbox.

The twins were altered with the standard TT4 becoming the TT6 with higher output generator, while the S8 was dropped. The S9 was continued and had the tank revised with a separate filler cap outside the toolbox in the top surface for greater convenience.

All the type numbers and little else changed for 1961 when there were detail alterations only. Model numbers went from C10, F10, U10, R10, TT6 and S9 to C11, F11, U11, R11, TT7 and S10

1959 Excelsior Talisman Twin, the TT4 with rear enclosure

Excelsior F11 Consort from 1961 fitted with Villiers 6F engine

67

respectively. Early in the year these machines were joined by three lower priced models based on existing ones but pared down a little to save costs. These were the EC11, ER11 and ETT7 with simpler enclosure panels for the battery and toolbox and similar cost saving moves.

There were further type number changes for 1962 with the C11, F11, ETT7 and S10 becoming the C12, F12, ETT8 and ETT9, but the only real alteration was to the 328 cc twin which assumed the same economy measures as the 243 cc model and lost its rear enclosure. The range was also considerably reduced and reflected the start of the decline of the English industry, with the EC11, U11, R11, ER11 and TT7 models all being dropped.

Early in 1962 Excelsior became one of the first firms to offer a kit bike and so save the cutomer purchase tax. The machine was a simple lightweight for road use, not the more normal competition, and was powered by the Excelsior 147 cc engine previously fitted in the Convoy and still used by the scooters. The cycle parts comprised a simple loop frame with swinging fork rear suspension and telescopics at the front.

The owner only had to assemble the major parts as engine unit, wheel and forks were already assembled, so the machine could be put together in an afternoon and at 99 guineas this gave a useful saving.

Excelsior continued this idea into 1963 but with an even further reduced range as the faithful rigid Consort was dropped along with the two twins to just leave two models both available complete or in kit form. One was the de luxe Consort which changed its type number to C14 and the other the Universal which again changed number to U14 but more significantly also changed its engine to the Villiers 31C. There were modifications to the fuel tank which became deeper and slimmer, while the separate tool boxes were replaced by a mid-section enclosure incorporating them.

The two models continued in this form into 1964 but the days of motorcycle production were coming to an end and late that year the Consort went to be followed by the Universal in the next year.

The company passed into the hands of the Britax organisation and in 1979 they re-entered the motorcycle field. This was with a miniature fold-up machine built in Italy harking back to Corgi days in its ability to be stowed away in the back of a car.

It was an interesting throw-back and gave about the same performance from its 50 cc engine, but the days of Excelsior motorcycles had gone.

Colours

1946: **O**—maroon including rims, tank lined; chrome plated exhaust pipe, black silencer.

1947: **O**, **V1**, **G2**—maroon, cream tank panels, cream engine panel on autocycles.

1948: **O**—as 1947. **V1**, **G2**, **S1**—black and cream in 1947 style, chrome plated rims.

1949: **U1**, **U2**, **R1**, **R2**, **M1**, **M2**: maroon with cream tank panel, painted rims, chrome plated exhaust systems.

1950/51: All motorcycles and **G2**—as 1949 in maroon with cream panels on tank and engine shield for **G2**. **S1**—black with cream panels on tank and engine shield; silencer painted.

1952: **G2**—as **S1**. All others: as 1950. **STT1**—beige with tank lined red.

1953: All models—as 1952, option for twins in polychromatic blue with cream tank panels. **F4**—black with twin gold lines on tank, aluminised finish for rims, chrome plated exhaust pipe, painted silencer.

1954: Existing models—as 1953, **C2** also in black with gold lined tank. **D12**—as **F4**. Other new models: maroon and cream.

1955: **S1**, **G2**, **F4**, **STT2**, **SE-STT2**—as before. **R4**, **R5**, **R6**, **C3**, **TT2**, **TT3**: black with gold lined tank, chrome plated wheel rims. **TT2**—during year also in black with red tank panel.

1956: **S1** and **G2**—pearl grey, painted silencer.

F4, **F4S**, **R6**—black with gold lined tank. **C1**, **C3**, **STT4**—cactus green. **A9** and **TT3**—pearl grey. All coloured models: black option. **C4**—pearl grey.

1957: All models—cactus green. Twins—chrome plated wheel rims. **F4**, **F6**, **SB1**—aluminised finish for wheel rims and hubs. **C4**—aluminised finish for wheel rims and silencer.

1958: All models—bronze green of deeper shade than 1957 colour, details as 1957.

1959: All models—cherry red, bronze green or two-tone pearl grey and red. **S9**—chrome plated tank top toolbox lid. **F4F**—dark green, tank gold lined, silencer green.

1960: **F10**—bronze green. **R10**—cherry red or two-tone pearl grey and red. All other models—as 1959. **C10**—chrome plated tank top strip.

1961: **F11**—grey with red headlamp shell and tank retaining strap. **C11**, **U11**—cherry red, Pampas green or dove grey. **R11**, **TT7**—Pampas green, dove grey or two-tone polychromatic red and dove grey. **S10**—cherry red or two-tone as **TT7**. **ETT7**—two-tone in pearl grey and peacock blue, dualseat with blue top and grey sides.

1962: All models—two-tone cherry red and grey or peacock blue and grey, the second standard for the **C12**. **U12**—peacock blue tank with ivory flash, peacock blue mudguards, upper front fork, chainguard; ivory frame, toolbox, battery box, lower front fork; dualseat with blue top and ivory sides.

1963/64: **C14** and **U14**—peacock blue and ivory as 1962.

Firefly

This was a frame kit devised in the late 1960s by Phil Jones and Brian Newberry to take all the Villiers competition engines and with minor modifications the Husqvarna and Greeves Challenger units.

There was nothing outstandingly special about the machine but it was well made and did incorporate a good number of useful ideas. One of the design aims was easy access and maintenance which was achieved by simple well engineered details.

The front forks were Bultaco, rear units Armstrong, and wheels from British Hub. Rear chain adjustment was by eccentrics at the wheel spindle, while the ignition coil was protected from vibration by being packed in sponge and pushed into a location in the air cleaner box.

The footrests could be fitted in high or low positions and easily altered to suit the day, while the tank was kept small to reduce top hamper to a minimum.

An enterprising project which like so many of the period survived for a while in a small way until killed off by VAT.

FLM

The initials stood for Frank Leach Manufacturing of Leeds and, unlike most small firms wishing to break into the lightweight motorcycle market, they used a JAP engine not a Villiers. Their frame construction was also out of the ordinary as the 125 cc unit was installed in a frame constructed of channel section steel with swinging fork rear suspension.

The control of this was by four springs housed

The 1952 FLM with 125 cc JAP engine unit in channel section frame

in a tubular cover beneath the engine unit with two connected to work in tension and the other two in compression. The fork arms turned down at the pivot point to link to these springs. Front suspension was by telescopic forks and the machine was well fitted out with a dualseat, well valanced mudguards and 5 in. diameter brakes.

During 1952 a second model fitted with a 197 cc Villiers 6E unit was on test and this model had a tubular frame, although it retained the under-engine springs for the rear suspension. No more was seen of this and the 1953 range when announced continued just with the 125 cc model but made available with rectified lighting as well as direct. Colours were given as beige and maroon, or gunmetal relieved with cream, and one change was to the footrests whose material became steel in place of the light alloy used at first. The centre stand was also strengthened.

Production of the machine was dependent on supplies of the JAP engine and as these dried up in 1952 the FLM disappeared from the lists the following year.

Francis-Barnett

Gordon Francis and Arthur Barnett were related by marriage before they joined names and formed their company in 1919. They used proprietary engines such as JAP, Blackburn or Villiers and in 1923 introduced their 'built like a bridge' frame constructed from a number of tubes bolted together. It was very cheap and easy to make and in one test two men assembled a machine in 20 minutes. Its strength came from the use of straight tubing in most instances plus **triangulation of the assembly which made it very rigid.**

In the 1930s the firm produced the Cruiser with a high degree of enclosure as well as built-in legshields, and in 1939 the Powerbike autocycle and 122 cc Snipe went into the range. There were plans for the army to use the latter but the Birmingham factory was then badly bombed.

Immediately the war ended the Powerbike reappeared and this was of typical autocycle construction. The engine was the Junior de Luxe and the rigid frame had a dropped top tube with the fuel tank hung below it and its lines continued down in side panels which enclosed the engine. Light blade girder forks were fitted along with hub brakes, pedalling gear, direct lighting and a bulb horn.

In the middle of 1946 it was joined by the Merlin 51 of 122 cc which housed its 9D engine in a rigid frame with girder forks. Twin exhausts and silencers were used and a neat touch was a three-pint oil tank fitted between the left chainstays. This was equipped with tap and filler cap with measure so that the rider could be independent of garages for his lubricant. The tank was matched on the right by a toolbox. Other fittings included speedometer, rear stand, saddle, rear carrier and direct lighting. Part of the rear mudguard hinged up to allow the wheel to roll out easily.

In June 1947 the company was amalgamated with the AMC firm, as was the James company later on, and in time the two came closer together until in 1962 Francis-Barnett production was transferred to the James factory and after that there was a steady move to badge engineering of identical models with only colour variations.

Before that happened the firm of Francis-Barnett continued to be in the forefront of producing fresh ideas in the lightweight class and they built up a reputation of providing a quality machine with many rider features.

For 1948 the Powerbike 50 had its forks changed to tubular girders with rubber band suspension, but otherwise the two models remained as they were.

The 1949 range brought changes with revised 122 and new 197 cc models. Both capacities were available with direct or rectified lighting and all machines had rigid frames and simple

1948 Francis-Barnett Powerbike 50 with Junior De Luxe engine. Typical autocycle

122 cc 9D engine unit of 1946 Francis-Barnett model 51 with twin exhausts

Falcon 55 from the 1950 Francis-Barnett range with 6E and rectified lighting

Falcon 58 from 1952 with swinging fork frame and 6E engine

telescopic front forks. The general equipment remained as on the Merlin 51 and the engines were the 10D for the Merlin 52 and 53 with rectifier, and the 6E for the Falcon 54 and rectified 55. All had a detachable rear number plate to assist wheel removal.

The Powerbike 50 continued with a bracing stay added to each front fork leg but was replaced by the model 56 in June 1949. This was powered by the 2F engine and had a revised loop frame to suit its installation. The fuel tank and engine shields were redesigned to suit the new frame but in general outline the model remained in the traditional autocycle form.

There was a slight reduction in gear ratios for the motorcycles during 1949 but in other respects they and the autocycle continued into 1950 without major change. One of the few detail alterations was to the saddle which was made adjustable for height and tilt.

With production being the order of the day there were only detail changes to the control levers and a revised and longer silencer for 1951. One feature that went was the spare oil tank on the left chainstay, although it did remain as an extra for export models.

71

Part Two Machines

Above **Francis-Barnett Kestrel 66 from 1954 fitted with 13D engine**

Top **Competition Falcon 60 Francis-Barnett of 1952**

Right **1956 Plover 73 with rear enclosure and alloy silencer beneath engine**

72

In August 1951 two new models joined the Francis-Barnett range, these being the Merlin 57 and Falcon 58 each with its Villiers engine in a new frame with swinging fork rear suspension. The frame was based on a single main loop with bolted on sub-frame and a fork that pivoted in rubber bushes pulled into place by a long through stud. The suspension units were made by the factory. Front forks were telescopic and most of the details as before but modified as needed to suit the new frame. Engines were the 10D and 6E and rectified lighting was fitted to both machines.

These two newcomers and the existing four motorcycles and one autocycle continued on into 1952 with two further machines for trials use. These were the Merlin 59 and Falcon 60 which used competition versions of the Villiers engines in suitably modified cycle parts. Frames were rigid, the front forks had stiffer springs, ground clearance was increased, saddle and bars raised, front mudguard sprung and silencer upswept.

During the year the Powerbike was dropped and when the 1953 range was announced the models 59 and 60 had also gone. During the year two series of models were built, Series 1 with a frame prefix beginning with the letter 'T' and fitted with the 10D or 6E engine, and Series 2 with part prefix 'TT' and using the 12D or 8E engine.

They were joined by a Merlin 61 trials model, a Merlin 63 scrambler, a Falcon 62 trials model and a Falcon 64 scrambler, the Merlins being fitted with 10D competition engines and the Falcons with the 7E unit. In both cases the trials machines had rigid frames and wide ratio four speed gearboxes, while the scramblers had swinging fork rear suspension and the close ratio gearbox.

During 1953 Francis-Barnett came out with what must be considered an early trail bike. This was the Falcon 65 Overseas model which came about when foreign riders bought the competition Falcon 62 and modified it a little for regular on and off road use. The Francis-Barnett result used the model 58 frame with stiffened front forks, the 8E engine and suitable features from the 62 to give it a trail bike specification.

In the autumn of the year all the Merlin models were discontinued and replaced by a single economy road machine, the Kestrel 66, fitted with the 13D engine. At the same time the Falcon 54, 55 and 58 went to be replaced by the 67, while the 62, 64 and 65 continued.

The Kestrel was a very simple machine with plunger rear suspension and direct lighting but it did share the same front forks of the range which had three rate springs. The rear units of the swinging fork frames had hydraulic damping added, while there were other detail improvements. These concerned the centre stand which became easier to use, the addition of a lifting handle, movement of the electric horn to a position just behind the headlamp, the option of a dualseat in place of the standard saddle, and a bigger tank on the Falcon 67 model.

Soon after the main 1954 range had been announced Francis-Barnett introduced a new and larger machine to augment it. This was the Cruiser fitted with the 1H engine of 225 cc and, while it looked similar to the road Falcon with swinging fork and telescopics, the frame was totally new.

It was based on a massive tapered down member formed from two pressings welded together and to the steering head. At the bottom a further pressing formed the front engine mounting. To the bottom of this were bolted a pair of tubes which ran out and back under the engine to curve up to the rear unit tops. Welded on plates carried these and the fork pivot. The two tubes were then bent to run forward as seat and tank rails to bolt to the down tube just below the headstock. Into this assembly went rear engine plates and to these were connected a single large pressing which ran up to the seat base and enclosed battery, tools and rectifier, while also forming part of the rear mudguard. With a dualseat as standard it gave a very sleek appearance.

The remainder of the machine followed normal Francis-Barnett lines but the brakes and rear tyre section were bigger and the model was fully equipped.

During the later part of 1954 the overseas model 65 was dropped and when the 1955 range was announced only the 62 trials Falcon continued. Like the other four models it was fitted with a new hydraulically damped front fork but no other changes. The scrambles machine became the model 72 still fitted with the 7E engine but having a strengthened and tri-angulated sub frame in addition to the new front forks.

The Cruiser became the model 71 and received full width hubs, while the Falcon became the 70 and gained a new frame, a dualseat and central enclosure. This sat just under the seat with side lids to give access to the battery, rectifier and tools, and also had the horn built into the front surface.

The Kestrel received three changes, the forks, an increase in engine capacity to 147 cc thanks to the three speed 30C unit, and full width hubs for both wheels. For 1956 it became the model 73 Plover and while it retained the same engine unit, the frame gained swinging fork rear suspension and the machine very substantial rear enclosure. By placing the undamped suspension springs well forward on the fork arms they became covered by the mudguard section of the enclosure and were thus out of sight.

1956 also brought 18 in. wheels for the Falcon which became the model 74, and an 18 in. front for the Cruiser. This was given model number 75 and fitted with a special centre stand. The stand went down in the normal way but was made easy to use as it left the wheels on the ground. A second lever on the right could then be depressed to raise the machine by cam action.

On the competition front the new numbers were Falcon 76 for trials and 77 for scrambles, both having the four-speed 7E engine in a new frame with swinging fork rear suspension.

Above **1956 trials model 76 with 7E engine**

Top **Francis-Barnett Falcon 74 with 8E engine from 1956**

Centre **Cruiser 75 with 1H engine built from 1955–57**

These two along with the road Falcon and Cruiser continued on for 1957, while the Plover changed its number to 78. In that form it gained a new dualseat, an electric horn mounted in the front of the centre enclosure and a tail silencer to augment the one mounted across the frame beneath the gearbox. The five models were joined by another for 1957, this being the Cruiser 80 fitted with the 249 cc AMC engine. In other respects it continued the features of the Cruiser 75 with Villiers 1H engine, except for the petrol tank which had the filler cap set beneath a lift-up cover, the main frame down tube which became a single pressing, a longer silencer on the right, 18 in. rear wheel, twin ignition and lighting switches in the headlamp shell, and the horn sited below the headlight.

1958 brought a reduction in the range with the Falcon models 74, 76 and 77 all coming to a stop along with the Cruiser 75, while the Plover 78 and Cruiser 80 continued. They were joined by a new Falcon, the 81, fitted with the Villiers 10E engine, this being the 9E type unit with vertical cylinder. The existing machines received detail changes only. Early in the new year the three road models were joined by a Cruiser 82 scrambler which used the 249 cc AMC engine in a swinging fork frame fitted with telescopic forks. The engine was tuned and the gearbox fitted with close ratios, while the machine was equipped solely for its competition use.

In May the road model range was extended further by the introduction of the Light Cruiser 79 fitted with the 171 cc AMC engine. The cycle parts followed the lines of the larger Cruiser and the centre section was enclosed by panelling that ran from the seat base down along the sub frame tube to the back of the gearbox and from the seat nose to the cylinder above the carburettor which was out of sight. The rest of the machine followed normal lines.

Aside from a colour option the complete range of five models continued for 1959 and were joined by a further competition model, the Trials

Above **Bob Currie testing the 1961 Francis-Barnett Cruiser 80 with its 249 cc AMC engine**

Below **1957–59 Falcon 81 model with the 10E engine with vertical cylinder**

Part Two Machines

The works at play with the 249 cc AMC powered scrambles model 82

83. This used the same frame as the scrambler and a lower compression ratio version of the competition 249 cc AMC engine with a set of wide gears fitted. Trials tyres and a saddle were fitted along with a bulb horn and a tubular toolbox under the seat. The silencer was of a substantial size and ran behind the right rear spring unit which was spaced out to accommodate it.

This range was joined by another Cruiser in November 1958, the 84 designed to provide maximum weather protection rather as the pre-war model. To this end the whole of the rear of the machine was enclosed from behind the engine cylinder. Tools and electrical equipment went under the seat which was easily removed for access, while the rear number plate detached to aid rear wheel changing as always. The model was also fitted with legshields as standard and a substantial front mudguard, while the pillion rests folded flush into the sides of the rear panels. Under the enclosure was basically the Cruiser 80 with 249 cc AMC engine but with the frame modified to accept the panels.

There were more changes for 1960 with the Plover 78, Falcon 81 and Trials 83 all being replaced, the first two by machines fitted with AMC engines of the same sizes. These were of 149 cc for the Plover 86 and 199 cc for the Falcon 87 both of which continued with the same basic cycle parts, those of the 86 as for the 78 while the 87 had a new frame of similar form to the earlier model. The new Trials model was the 85 which had a padded seat to lower the riding position and a smaller front brake. A direct lighting set was an option and connected via a multi-pin socket. Models 79, 80, 82 and 84 all continued in the range.

Having done so much revision Francis-Barnett decided to leave well alone for 1961 and the only change was the deletion of the model 79 from the range. All models were now powered by AMC units whose assembly was transferred to the Villiers works during the year. The six machines carried on into 1962 as well with the only change of note being the removal of the legshields and crash bar from the enclosed Cruiser 84. The more standard 80 gained the option of rear chain enclosure.

The range was augmented with two further machines for 1962. The first to be announced was the Cruiser Twin 89 which had the Villiers 2T engine housed in the model 80 cycle parts and the choice of three colour finishes.

The second was much more striking and was the Fulmar 88 styled with panelling that ran from the headlamp back under the seat to the rear mudguard. Beneath this went a spine frame built up from tubes with a degree of bolted construction and hung from this was the 149 cc AMC engine, built then by Villiers.

Front suspension was by short leading links with enclosed units and a conventional swinging fork looked after the rear wheel. The panelling was for style and carried a toolbox in the dummy tank top, while the tank itself was a box set in the frame with the filler cap beneath the hinged dualseat. The handlebars were covered by a pressing.

The Fulmar continued on into 1963 along with the Falcon 87, Cruiser 80 and Cruiser Twin 89 with only detail changes, but models 82, 84, 85 and 86 were all dropped. The four road machines were joined by two sports models, one the Fulmar 90 and the other the Cruiser Twin 91. Both featured sports handlebars, a flyscreen and polished mudguards, while the Fulmar also had a four-speed gearbox and special silencer and the Twin a higher compression ratio and smaller styled tank.

Also new for 1963 were a pair of 246 cc competition machines, the Trials 92 with Villiers 32A engine and the Scrambler 93 which had a Villiers 36A unit fitted with a Parkinson square barrel conversion. Both machines had four-speed gearboxes with wide ratios for trials and close for scrambles, while the frames were fitted

Above **1960 model 87 Falcon with 199 cc AMC engine**

Top **Francis-Barnett Cruiser 84 with full rear enclosure**

Centre **Trials model 85 built from 1959–62 using 249 cc AMC engine**

Part Two Machines

Above **1962 Francis-Barnett 149 cc Fulmar 88 with spine type frame and AMC engine**

Below **1964 Cruiser 89 with 249 cc Villiers 4T engine**

Below **Sports Fulmar 90 from 1963 with Villiers/AMC engine and eye catching bodywork**

with Norton telescopic front forks and swinging fork rear suspension. The trials model had 5 in. brakes and the option of lighting, while the scrambler had a 6 in. front brake, a smaller fuel tank and no lights.

With times for the industry becoming harder there were few changes for 1964 with models 87, 88, 90, 92 and 93 all continuing while the Cruiser 80 was dropped. The two Cruiser Twins had an engine change to the 4T and there were two new models.

The first was the 149 cc Plover which re-appeared as the model 95 in a single tube spine frame with massive dualseat and short, humped back tank. The second model was the Starmaker Scrambler 94 which comprised the model 93 fitted with the Starmaker engine in place of the 36A. Early in the year the 93 was dropped.

This left a total of eight models which continued into 1965 without change, with five of them running on into 1966 as well. Most had counterparts in the James range for since the move by Francis-Barnett in 1962 to the Greet factory the machines had become more common. Thus the Sports Twin 91, Trials 92 and Scrambler 94 all had their James counterparts as did the new low-cost 149 cc model 96 which replaced the Fulmar and Plover models which were dropped. The 96 used the AMC engine in a simple loop frame of conventional design. Models that remained unique to Francis-Barnett were the Cruiser Twin 89 and Falcon 87.

Time was however running out for the long established make and, in October 1966, production of Francis-Barnett motorcycles ceased.

Colours
1946: **50** and **51**—black with gold lining for tank and autocycle side panels.
1948: **50**—as 1946 except wheel rims chrome plated. **51**—as 1946.
1949/50: **50**—as 1948. **52**, **53**, **54**, **55**—as **51**. **56**—black with gold lining for tank and side panels; chrome plated wheel rims.
1951: **52**, **53**, **54**, **55**—as before plus option of azure blue. **56**—as 1949. **57**, **58**—as **52** to **55**. **59**, **60**—black with gold lining.
1952: As 1951.
1953: As 1951; for export models chrome plated tank panels and wheel rims available. **61**, **62**, **63**, **64**, **65**—black with gold lining.
1954: **66**—azure blue with gold lined tank and new F-B monogram transfer in red, gold and silver; matt silver wheel rims. **67**—azure blue or black with gold lined tank, plastic tank badge in red, gold and silver; painted rims. **62**, **64**, **65**—as 1953. **68**—dark green, twin gold lines on tank, plastic tank badge; green dualseat; chrome plated wheel rims.
1955: **69**—black, tank nose and panels in yellow lined gold; matt silver wheel rims. **70**—black with gold tank lining; chrome plated wheel rims; optional chromed plated tank panels. **71**—as **68**, wheel rims chrome plated with green centres lined gold; optional chrome plate tank panels. **62** and **72**—black, gold lined tank, black rims.
1956: **73**—Arden green, twin gold lines on tank; matt silver wheel rims and hub end covers. **74**, **75**—Arden green, twin gold lines on tank; chrome plated wheel rims, matt silver hub end covers. **76**, **77**—Arden green; tank dull chrome plated; alloy mudguards.
1957: **78**—as **73** except cadmium plated wheel rims. **74**, **75**, **76**, **77**—as 1956. **80**—as **75**.
1958: **78**, **80**—as 1957. **81**—as **74**. **82**—as **76** and **77**; tank lined green. **79**—Arden green; tank and side covers lined white; chrome plated wheel rims lined green and gold.
1959: **78**, **79**, **80**, **81**, **82**—as 1958. **83**—as **82**. **78** option—tank panels only in Dover white. **79**, **80**, **81** option—tank, mudguards, enclosure in Dover white; toolbox lids white for **81**, green for **80**. **84**—Arden green frame, forks, legshields, rear mudguard; Dover white front mudguard and rear enclosure panels which had a green flash on each side carrying the model name in white; green seat; chrome plated wheel rims, headlamp rim, safety bar supporting tops of legshields and

Francis-Barnett 1966 Sports Cruiser 91 with 4T engine and styled tank

1964 Plover 95 with 149 cc AMC engine built by Villiers

lifting handles; option of chrome plated petrol tank.
1960: **80**, **82**, **84**—as 1959. **85**—as model **83**. **86**—Dover white for tank panels with either Arden green (as model **78**) or Burma red. **79**—as 1959, option in green and white or with Burma red in place of white to give green and red; option of all green with narrow horizontal white bands on tank and side panels. **87**—green with white bands as **79**, or as model **81** option.
1961: As 1960.
1962: **82**, **85**—as 1960. **80**—tank lower half and rear mudguard white with remainder Arden green or black; option of all Arden green with white tank flash. **87**—as 80 plus option of black and Tartan red. **84**—Arden green, Tartan red or black with white for front mudguard, lower tank half and lower part of enclosure panels. **86**—as 1960 in white with Arden green, or Tartan red. **89**—Arden green or Arden green with rear mudguard, mid-section enclosure and lower part of fuel tank white, or in black and white to the same combination. **88**—Dover white panelling and rear mudguard with frame, forks and front mudguard in black or Tartan red.
1963: **88**—as 1962, or in Arden green for panelling and front mudguard with frame, forks and rear mudguard silver. **80**, **87**, **89**—green, or green and white in 1962 style **87** with side panels in white upper and green lower sections.
1964: **87**, **89**—Arden green with white rear mudguard, upper side panels and lower tank panels. **88**—green and silver, or Tartan red and white. **90**—bright red and silver with gold lining; chrome plated mudguards and chainguard. **91**—red and silver. **95**—Arden green with white tank panels.
1965: **89**—Arden green with gold lining, or in green with cream rear mudguard, centre panel and lower tank panel. **91**—red forks, frame and tank; alloy mudguards; chrome plated tool and battery box lids; silver tank side panels. **95**—as 1964.
1966: **87**—Arden green and white as in 1964. **89**—As 1965 in green and broken white. **91**—As 1965. **92**, **94**—Arden green and silver. **96**—metallic green tank, centre section and mudguards with black frame and forks.

Greeves

The name of the machine was that of the managing director of the firm that built the prototype in 1951—Bert Greeves. The firm was Invacar who up to then had specialised in the production of invalid carriages powered by small Villiers engines and the man who was to become the Greeves sales director, Derry Preston-Cobb, was himself confined to one of their three-wheelers but no less an enthusiast for all that.

Two prototypes were seen in May 1951, both of unusual appearance due to their suspension systems. Both wheels were sprung and each used a pair of Metalastic torsion rubber elements as the spring medium. The frame itself was tubular with a cross-over at the headstock but the front forks were trailing link with the torsion bushes mounted at the pivot points. The appearance was made more unusual with a tubular loop around the front of the wheel and a long, rear-pointing, torque arm on the brake backplate which was connected to the fork tube by a link which thus worked in compression.

At the rear the wheel was carried in a swinging fork which was linked on each side to a tube running from the wheel mounting up to the levers attached to the torsion bushes mounted under the seat. Power unit was a 197 cc Villiers and the machine was fully equipped but looked a little odd. The second prototype looked more odd as it was set up for competition and the skimpy tank and mudguards accentuated the functional array of tubes used for frame and forks.

Development continued for the rest of 1951 and through 1952 as well, although there were no problems with the rubber suspension units as these were in use in the invalid carriages so were well proven.

In the end it was September 1953 before the production machines were announced and, although their appearance had become more conventional, they bristled with unusual features. The frame was unique as it was part tubular and part alloy casting. The single top tube was welded to the steering head and this part then went into the casting mould. The result was a massive I-section beam that tapered down from the headstock to the front of the engine and which also was cast around the top tube, so irrevocably locking tube and casting together.

To the bottom of the down beam were bolted two further deep I-section members which ran under the engine unit to support it. They were attached by a pair of well spaced bolts and carried the footrests while also providing a location for the rear fork pivot behind the gearbox. To them, on their outer side, was bolted a tubular subframe which ran up, bent round housings for the rubber torsion suspension units and then in to bolt to the back end of the top tube. Lugs on the tube supported the outer end of the fork pivot so this was located at four points across the machine.

The rear fork was built up from a U-shaped member in round and tapered tubing welded to a pair of supports in square section. These carried the bonded rubber pivot bushes which were clamped up by a single spindle. The wheel fork ends carried mountings for struts which ran up to the suspension units which incorporated adjustable friction dampers. Conical snubber blocks limited fork travel.

At the front, leading link forks were employed retaining the torsion rubber suspension units at the pivot point with friction dampers. The links were bridged by a rear loop and the main tubes joined by the bottom crown, that at the top clamping to them.

This frame was used for two road and two competition models all fitted with the Villiers 197 cc engine unit, plus a twin powered by the 242 cc Anzani one. The road machines were fitted with well valanced mudguards, a dualseat, small tank and headlamp shell mounted on four small stays. The seat and rear mudguard was combined into one unit which could be removed by releasing three bolts thus making rear wheel removal easy. The cylindrical toolbox, battery carrier, horn and rectifier were all tucked neatly away in the area under the seat nose.

The competition models had the equipment to suit their use but included one further surprise with plain bearing hubs, found to withstand the rigours of scrambling in wet and dirty conditions much better than ball races with less than perfect sealing. In all cases 6 in. diameter brakes were fitted, unusually large for a lightweight.

While all the single engines were the Villiers 8E, the Standard model only had three speeds while the others had four with close or wide ratios to suit their purpose. The twin with its Anzani unit was called the model 25D Fleetwing and copied the singles in equipment and general specification.

These five models comprised the range for 1954 but for 1955 there were some changes with the addition of the larger capacity 32D Fleetmaster fitted with the 322 cc Anzani twin. General changes concerned the use of well sealed ball race journals for all wheels, polished cast alloy tank motifs, keep plates to add to the front wheel spindle security, an improved run of front brake cable external to fork tube instead of within the right one, and mounting brackets for the road model headlamp welded to the fork tubes.

New models were the standard 242 cc 25R Twin and 197 cc R3 and R4 with an all tube frame. In this the assembly of top tube, cast beam and cast engine cradle plates were replaced by a simple open loop tube which bolted to the subframe which in these cases supported hydraulically damped spring units of conventional type. These were also fitted to the scrambler but the other models retained the rubber in torsion as did the leading link front forks.

Most models retained the 6 in. brakes but the Fleetmaster had twin 6 in. at the front and a 7 in. at the rear, while the R3 and R4 had 5 in. for both

Above **Greeves road model, the 1956 20D with 9E engine at Southend Airport**

Below **1956 model R3 with 8E unit and tubular frame, usual Greeves front forks**

Above Road model 25D in 1958 with Villiers 2T engine and familiar cast beam frame

Top 1957 scrambles Greeves model 20S with 9E engine and stub exhaust

Right 1956–57 trials model 20T with 197 cc 9E engine

Part Two Machines

Above **Greeves model 20TA Scottish Trials in 1958 with 9E engine**

Above **1959 model 24SAS Hawkstone Special Scrambler with 31A engine**

Below **A 250 Greeves in ISDT form with air bottle, spare cables and special barrel**

wheels. The trials model had a 4 in. section rear tyre fitted, footrests moved back, and increased steering lock.

For 1956 there was a general change to the 9E engine for the single cylinder machines except in the case of the 20R which continued with the 8E. The de luxe 242 cc twin was dropped but the standard one continued along with the larger Fleetmaster.

Changes included a new dualseat, a toolbox beneath it which could be detached from the machine when needed, a valanced rear number plate, a more forward mounted and larger fuel tank and a headlamp cowl for some models. The competition models gained a seat pillar bolted into their frames and the trials one had a saddle fitted as standard. All frames had conventional rear suspension units so no more was seen of the torsion rubbers at the rear, although they continued to be used in the front forks.

Signs that the Anzani engines might not be available for too long possibly influenced the appearance of the 25D Fleetwing fitted with the Villiers 2T engine for 1957. Also new on this, the 32D and the competition models was a new version of the leading link fork with Girling hydraulic dampers hidden in the fork tubes to keep control of the torsion rubber suspension units. The 25D was fitted with a side battery cover which was also used on the 20D Fleetstar and the Fleetmaster. Both the larger twins were fitted with larger front wheels, while the competition models had theirs go down a little, the result for all four machines being a 20 in. diameter rim. All models in the range had the front brake lever reversed to improve the efficiency and this resulted in a revised cable run.

During the middle of the year the economy model with the 8E engine was replaced by a Sports version, the 20R4, fitted with a 7E unit with four-speed close ratio gearbox. This only remained in the lists for a few months.

When the 1958 range was announced it was seen that the predictions with the Anzani engines had come true and both the twins with those power units were dropped. This left the two D road models with 9E and 2T units which continued with detail changes to the front mudguard valances and the tank nameplates which had the model name added below that of Greeves.

On the competition front the company began what was to be a whole series of models for trials and scrambles in two engine sizes, 197 cc using the 9E unit and 246 cc with the Villiers A series. The models were to be given a series letter over the years which in time ran from A to J in one case, and were coded 20 for the 197 and 24 for the 246. The latter did not join the range until late in the year and was preceded by the code 25 models with the 2T engine. The code used the number giving the engine size followed by a letter sequence which began with T for trials, S for scrambles, and D for the occasional road model. This was followed by the series letter and sometimes an added S for special. Later M was also used for moto-cross and S for Silverstone.

Of more interest for 1958 was a new frame for the competition machines with a much stiffer cast down beam and engine cradle plates that gave an extra 3 in. ground clearance at the front decreasing to 1 in. at the footrests. The rear subframe and fork became narrower so the whole effect was a real improvement over the earlier model. The machines were called the 20TA Scottish Trials and 20SA Hawkstone Scrambler.

At the end of 1957 these were joined by the 20SAS Hawkstone Special Scrambler which had a specially tuned engine, and the 25TA and 25SA which were trials and scrambles machines fitted with the 2T engine unit. These did not last long for two-stroke twins were never very good off road, and the 25TA went when the 1959 range was announced while the 25SA only continued for another year to special order. Of the road twins the 25D was replaced by the 25DB Sports Twin.

85

Part Two Machines

Above **The 1963 Greeves Silverstone 24RAS road racer, start of a line**

Top **1961 Greeves 32DC road model twin with radial fins on brake drums**

Above **Strangely styled 25DCX model of 1962–63 with 2T engine. Handlebar fairing and spats for the forks not really in Greeves' image**

Top **1961 model 24TES with TDS exhaust**

The 246 cc machines were introduced for the 1959 season, all powered by the 31A engine tuned to suit its purpose, and numbered 24DB Sports Single, 24TAS Scottish Trials Special and 24SAS Hawkstone Special Scrambler. All three shared the cycle parts of the 197 cc models which continued in competition form only as the 20TA, 20TAS and 20SAS. The road 197 was dropped for a spell.

All models had the newer composite tube and cast alloy frame with leading link forks and the competition ones had a chain oiler in the left rear fork leg. The 246 cc scrambler came with an alloy, radially finned muff cast onto both brake drums, and this was listed as an option for the other machines. Other options for the competition models included direct lighting, road equipment for the SAS types, a larger tank and a

1965 Greeves Silverstone 24RCS with expansion chamber and perspex nose to fairing

dualseat for the trials machines fitted as standard with a saddle. Thus the Greeves sought to be all things to all riders.

The 250 twin and single cylinder road models continued with no apparent alteration for 1960, although the single changed its engine type to the 32A as did the trials model. This and the 197 cc trials machines all had a steeper head angle incorporated into steering geometry changes which offset the fork tubes forward of the steering column, increased the steering lock and reduced the wheelbase. Suspension rates were softened and the rear units lost their lower shrouds, while tank capacity went up a little.

In January 1960 new scrambles machines were introduced as the 20SCS with 9E engine and 24SCS with the 34A. They shared common cycle parts with a heavier section for the cast alloy frame beam to allow taper roller head races to be fitted. The rear subframe was also modified to increase its stiffness and allow the fitment of a 4 in. rear tyre. The 250 continued to be the only model fitted with the radial finned brake drums as standard.

These models were joined by a further road

1964 Challenger 24MX1 with Greeves engine, a functional moto-cross machine

machine during the year, this being the 32DB Sports Twin fitted with the 324 cc Villiers 3T engine in the same cycle parts as the 25DB. As was becoming usual these changed model numbers to 32DC and 25DC for 1961 and in the process were fitted with new frames based on those used by the scrambles machines with taper roller head races.

Also new for 1961 was the 197 cc road model which reappeared first as the 20DB and then became the 20DC, while the 246 cc version became the 24DC. The competition models became the 20TD and 24TDS for trials, while the 20SCS and 24SCS continued for scrambles. The last was joined by a significant newcomer, the 24MCS Moto-Cross Special, this having the normal cycle parts but an engine fitted with a special alloy cylinder head giving an 11:1 compression ratio and a square finned alloy barrel. The trials models were also modified with the engine unit moved forward and down and the rear fork pivot forward.

This did not stop Greeves from introducing a new frame for the 1962 trials models coded 24TE when fitted with the 32A engine, and 24TES when that engine was in turn fitted with a Greeves alloy head and barrel. As was the company practice, the new frame was really another step along the development path they had followed since 1953 and now the trials and scrambles frames were different units although both used the cast beam. Front suspension remained leading link.

Much of the rest of the range continued with little or no change although the tank fixing on the 24MCS altered to one front bolt and a pair of hook springs. Early in the year it became the 24MDS with an increased compression ratio and more power. The frame was modified slightly and full width hubs fitted, both the hubs and the

East Coaster road model from 1965–66 coded 25DC, fitted with 4T engine

backplates being in light alloy. The front backplate was linked to the fork tube by a torque arm to avoid any tendency for the front end to rise under heavy braking.

At the same time a 197 cc trials model 20TE was added to the range using the same cycle parts as the 24TE. In April 1962 two further models were introduced, both new versions of the twin cylinder road machines. These were the 25DCX and 32DCX Sportsmen which shared cycle parts and were fitted with the 2T and 3T engine units. They were styled by a handlebar fairing with small Perspex screen and by the addition of spats over the normal leading link forks. A tank with knee recesses was fitted, the footrests were positioned well back, and full width hubs used. Both machines had twin exhaust and silencer systems unlike the earlier twins which used a two-into-one arrangement.

During 1962 the company built some ISDT replicas by combining parts from road and competition models and for 1963 continued with modifications for several competition models to provide a variety of specifications to suit peoples' needs and pockets. One, strictly Greeves, change was to a cast iron liner in place of the chromium plated bore of the square alloy barrel. Another was a square section exhaust pipe curled tightly round that barrel to its small silencer on the trials model. On the 24TES the engine plates were boxed in for strength and to act as a crankcase shield and also pierced with holes to reduce weight.

Road models for 1963 were augmented by the 25DD and 32DD Essex twins which combined the fibreglass tank, fork shrouds and twin exhaust systems of the DCX models with the styling of the DC except for the mudguards which became plastic and well valanced. The other four twins

Greeves Silverstone for 1964, the 24RBS, a model that many used for their first race

continued as did the 197 single, but the 24DC was dropped. Also out was the last of the 197 and the older 246 scramble machines and the TD trials models, but the TE trials models in 197 and 246 cc sizes continued as did the 24MDS which was joined by the 24MD.

The real news for 1963 came at the end of October 1962 when a completely new form of competition Greeves was announced, the 24RAS Silverstone production racer. This came about due to the success in club racing of a scrambles Greeves modified as far as controls and tyres went, plus the interest of Bert Greeves himself.

The result was a slim road racer which used the 24MDS frame with pierced engine plates but reduced across the rear fork pivot. Front forks and rear suspension followed MDS practice but wheels, tank, seat, tyres and controls were all suited to road racing. The prototype engine was the 34A fitted with Greeves head and barrel fed by a 1·25 in. bore GP Amal. The expansion chamber ran at a high level alongside the barrel and the tailpipe tucked under the seat. It was an immediate success at club level which was what it was intended for.

By the time the 1963 season rolled round the engine type had become the 36A, also used for the moto-cross model, and the carburettor size had gone up to 1·375 in., while a rev-counter drive had appeared on the right end of the crankshaft. Power output was up to 30·5 at

The 1965 model 24TFS with Greeves Challenger top half on 32A bottom

7100 rpm, a useful increase on the original engine's 25 bhp. A head fairing was added and the exhaust pipe was remounted low down on the left.

Also new for 1963 was the 24ME, a scrambles machine fitted with the Villiers Starmaker engine, but this model was only produced for that one year. Again a few ISDT replicas were built based on the works machines.

For 1964 the two 197s continued, these being the 20DC and the 20TE, but the three 32 road machines were dropped. The 25DC carried on with its 2T engine but both the 25DCX and 25DD changed to the 4T unit. Models that continued were the 24TE, 24TES as the MKII, 24MD and 24MDS.

In February 1964 a new scrambler, the 24MX1 Challenger, was announced and followed shortly by a second version of the Silverstone, the 24RBS. Both were fitted with a completely new engine and the road racer had a five-speed gearbox.

The engine followed the logical development pattern that had begun with the 9E; moved on via 250 conversions to the A series, passed into the square barrel phase, and reached the limit of the Villiers bottom end. Thus the new unit was based on a Greeves designed and Alpha built flywheel assembly with caged roller big-end and ample main bearings. Bore and stroke remained at 66 × 72 mm. On top of the well ribbed crankcase went a larger alloy square barrel with

Greeves Anglian from 1966, 24TGS model with banana front forks, engine as 24TFS

A Greeves with telescopic forks and twin exhausts, the 1967 36MX4 Challenger

cast iron liner and onto that went a new cylinder head with central plug and fins designed by air flow tests to keep it as cool as possible. Compression ratio was 12:1 at first and the piston a conventional two-stroke type with Dykes top ring and plain second. The ratio was later reduced to 10:1 for the MX1 and 9·5 for the RBS, with carburettor sizes of 1·187 and 1·375 in. and down-draught angles of 8 and 22 degrees respectively. Two insulating pieces went between barrel and carburettor and between them went a steel support for the racer's GP float chamber, or an alloy deflector to keep hot air away from the Monobloc used on the scrambler.

Ignition was by a small Swedish Stefa magneto on the right end of the crankshaft, while a rev-counter was driven from the left. To the back of the crankcase was attached a four- or five-speed Albion gearbox and this was driven by a single strand primary chain.

The MX1 had this unit installed in cycle parts which were typical Greeves, although the engine plates just linked the front of the engine to the cast frame beam. Under the engine, gearbox protection was afforded by two tubular bars and a bash plate. Forks were leading link, tank and rear mudguard in fibreglass, front guard in a semi-flexible plastic, hubs full width light alloy and under the seat was tucked a large air filter on the left and the expansion box on the right.

The RBS had the five speed gearbox, the expansion box under the engine to emerge on the right, the same basic frame and forks as the Challenger and continued with its head fairing and racing seat. A steering damper and front mudguard were added and the front brake well ventilated with air scoop in the back plate and holes in the hub.

The Challenger engine was also used for the ISDT models built that year based on the MX1. In September the range for 1965 was announced and, while the 197 cc single and 25DD twin continued unaltered, the 25DC was fitted with the 4T engine and the 25DCX with head fairing was dropped. The 197 cc 20TE ran on but went during 1965 as did the 24TE and 24MD. The 24TES became the 24TFS with Challenger head and barrel on a 32A bottom half and received other detail changes to ensure a smooth underside to the machine. With the success of the MX1 the need for the 24MDS went so it was dropped.

At the end of 1964 some indications of future developments were seen, the first being a revised barrel selector for the gearbox fitted with the Challenger engine. Early in 1965 the works scramblers were shown to the press finished in pale pastel green in place of the familiar blue and fitted with revised leading link forks with the pivot point further back. Control was by a pair of Girling spring and damper units so for the first time Greeves had a front fork without Metalastic torsion rubber units. The new forks became known as the 'banana' type. The next rumour was of a 360 cc moto-cross machine, but Bert Greeves said this was 'rather premature'.

Also early in 1965 the Challenger was replaced by the 24MX2 and the Silverstone by the 24RCS, really just the earlier model fitted with a full road racing fairing. In the middle of the year the 25DC

The Silverstone in final form, the 24RES of 1967–68

East Coaster road machine appeared, powered by the 4T twin engine in a conventional set of cycle parts. Later in the year all the other 249 cc twins were dropped while the East Coaster only ran on to the middle of 1966 before it too went.

1966 also saw the end of the 197 cc road model, the sole remaining machine of that capacity in a rather depleted range that reflected the downturn of the industry. The three competition machines all changed their designation to 24RDS Silverstone, 24MX3 Challenger, and 24TGS with a new name, Anglian. Both off road models were fitted with the banana fork and the Anglian frame was tidied up and lightened. A return was made to the old style steel hub as this worked better in trials use, while the engine continued to be the 32A with Challenger parts. The Challenger engines themselves gained a duplex primary chain, this having been in use on some models during 1965. The Silverstone was fitted with a new clutch and gearbox, and modifications to engine, gear linkage and fairing for the new season.

For 1967 the competition models again changed their coding to 24THS, 24MX5 and 24RES and were joined by the rumoured 360 scrambler, the 36MX4. This was powered by a new Challenger engine with dimensions of 80 × 72 mm which gave it a capacity of 362 cc. While similar in design to the 250 it was all new and featured twin exhaust ports whose pipes connected into a single expansion chamber under the engine. Other features were much as on the smaller engine and the gearbox was similar but reduced in size with an improved selector mechanism.

The frame was new and continued the cast beam tradition with single top tube but this split to a duplex at its rear end, the pair of tubes curving down to the rear fork pivot. Front forks were the Greeves banana but all off-road models had the option of Ceriani telescopic forks if required. The hubs on the 360 were Greeves conical alloy units with 6 in. diameter brakes but 3 lb lighter than the full width type.

The 24MX5 continued with the improved gearbox, an Amal Concentric, restyled tank and longer wheelbase, while the 24THS also received a Concentric, a change to a 37A bottom half and a number of detail modifications to the controls. The 24RES gained a colour change, a new cylinder head, alternative footrest positions and a humped-back seat with suede cover. When the season came round the racer was also found to have a 7 in. twin leading-shoe front brake.

At the same time there were alterations to the production barrels for the 360 Challenger which became held down by eight studs and increased its fin area. The head fixing was also altered to six studs and this arrangement replaced the four through-bolts used up to then and continued on the 250.

For 1968 the Silverstone was joined by the 344 cc Oulton which used the 360 moto-cross bottom half with the stroke reduced to 68·5 mm combined with the common 80 mm bore. Twin exhaust pipes led to separate expansion boxes, at that time a common practice, while the frame and cycle parts copied the Silverstone. That model continued as the 24RES along with the 36MX4 which was joined by a 250 version based on a new short stroke engine with 70 × 64 mm dimensions and the same cycle parts. The engine construction duplicated that of the larger model and the machine was typed 24MX4. With its appearance the 24MX5 was dropped. In the trials field the 24THS was replaced by the 24TJS Anglian which had telescopic front forks fitted as standard. This was joined by the 24TJ Wessex which retained the banana leading link fork and was an economy version with steel tank and fitted with a 37A engine with iron barrel.

As Greeves built their own engines for many models they were less affected than many in the ailing British industry when the Villiers units were no longer available. However, a contraction was necessary and thus 1969 saw a move into kit bikes, a new name for the scrambles machines and an end to the Silverstone, the few Oultons built, the Anglian and the Wessex.

The new name was the Griffon, produced in two sizes using the short stroke 246 cc engine and a new one of 380 cc based on 82 × 72 mm dimensions and developed from the 362 cc engine. Both followed the design of the Challenger engines closely and had a Greeves designed four-speed gearbox and all metal clutch.

The frame was new and all tubular in Reynolds 531 with Greeves-Ceriani front forks and hydraulically damped rear units. Hubs were conical in light alloy with drum brakes. The 380 retained the twin exhaust ports whose pipes joined above

Greeves Challenger 36MX4 with 362 cc engine and Ceriani forks

the cylinder head to feed into a single expansion chamber mounted high on the right.

These machines were joined in 1970 by the trials Pathfinder fitted with a 169 cc Puch engine and this also had a tube frame and telescopics but the brakes were in full width, and smaller, hubs. It was joined by an enduro version in 1971.

1971 also saw the adoption of the 380 QUB engine in the Griffon, this being a unit incorporating developments suggested by Dr Blair of Queens University, Belfast. These affected port timing and expansion chamber but the most obvious was a change to a single exhaust port.

The two Griffons continued with small changes each year but the great days of the company had gone. In 1973 Bert Greeves retired and the firm passed to other hands. Relatively few machines were built and in 1977 the range was

Greeves Oulton, the 1968 model 35RFS based on the Silverstone

1969 Greeves Griffon with 250 cc engine, tubular frame and telescopics

discontinued although sales of general accessories had helped out. Rather sadly the Receiver had to close the company doors in 1979 and the flying blue two-stroke with its rubber suspension that Brian Stonebridge and Dave Bickers had campaigned so successfully was no more.

Colours
1953/54: **20R** standard road— black frame tubes, forks, mudguards, hubs, headlamp shell; grey wheel rims. **20D** de luxe road— as **20R** in blue. **25D** Fleetwing—in blue as **20D** but with chrome plated wheel rims. Competition—in blue with chrome plated wheel rims. In all cases alloy frame members left natural and Greeves motif on silver tank panel.
1955: All models—chrome plated wheel rims, matt on roadsters, polished with matt well for competition; Moorland blue finish as 1953; polished, cast alloy tank motif.
1956: All models—as 1955 in Moorland blue with chrome plated wheel rims polished with matt well except for **20R3** with all matt rims; light blue dualseats.
1957: Competition—fuel tank and dualseat in light blue with dark blue option. Road—dark blue frame, forks and mudguards with fuel tank, headlamp, shroud and battery case in pale blue-grey; dualseat with small blue checks.
1958: **20SAS** only—dull chrome fuel tank.
1959: Road—Moorland blue frame tubes, forks, headlamp shell, chainguard, centre panels and tank side panels; remainder of tank Essex grey; chrome plated wheel rims; polished alloy mudguards. Competition—as road except tank dull chrome plated.
1960: As 1959.
1961: As 1959 except **20DC** with all dark blue tank with gold lining and **32DC** in dark blue and silver with winged motif on side of centre panels.
1962: As 1961 except **24TE**, **24TES** and **24MCS** which had polished alloy tanks. **25DCX**, **32DCX**—pale turquoise blue fairing, tank, frame tubes, centre panels, forks; pale jasmine yellow flash on each side of fairing and tank panels; polished alloy mudguards; chrome plated wheel rims.
1963: As 1962.
1964: Generally as 1962 but **25DD** also available in peacock blue with tank in light and dark blue. **MX1**—Moorland blue frame and forks; white mudguards and tank.
1965: As 1964.

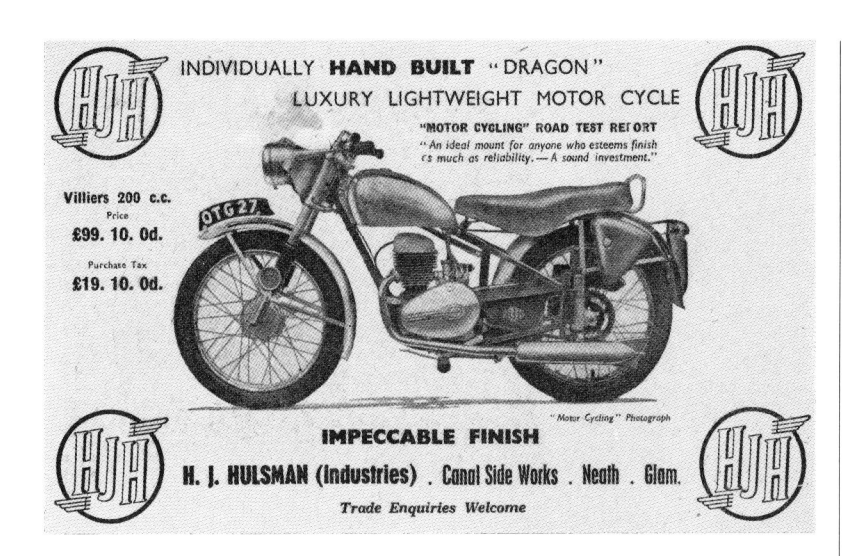

HJH magazine advert of 1955 showing the 197 cc Dragon

1966: **24TGS**—light grey frame and forks; red tank; alloy mudguards. **24MX3**—Essex green frame and forks; white tank and mudguards.

1967: **24THS**—silver frame and forks; red tank; alloy mudguards and chainguard. **24MX5**—Essex green frame and forks; light grey tank and air filter box. **36MX4**—silver frame and forks; hunting green tank, air filter box and front part of rear mudguard; front mudguard and rest of rear one in alloy. **24RES**—silver frame and forks; red tank; white fairing.

HJH

In the 1950s a number of businessmen saw the Villiers 8E as a means to becoming a motorcycle manufacturer and one result came from the works of H.J. Hulsman (Industries) of Canal Road, Neath, Glamorgan in Wales. As befitted a Welsh machine, it was given the model name Dragon and was neat if conventional in appearance. Like the DMW it used square section tubing for the frame which had plunger rear suspension and telescopic front when first seen in 1954. It was well finished and came equipped with dualseat, pillion rests and twin toolboxes which set it off well. The brakes were not perhaps as large as were sometimes needed in the Welsh hills but worked well enough when cruising.

In 1955 the Dragon was joined by the Super Dragon fitted with Earles forks and a trials model which used the four-speed 7E engine in a rigid frame with Earles forks. Further new models were the Dragon Major which had a 1H engine in a swinging fork frame fitted with Earles forks and the Dragonette, a smaller model using the 30C engine in a rigid frame with telescopic front forks.

At the end of the year the Super Dragon was dropped and the Dragon became the base 200 cc model with the rigid frame and teles of the Dragonette which continued and was joined by a Sports version with swinging fork frame. The trials model also gained this form of rear suspension and four other new models were listed.

These were all of 200 cc and comprised the Sports Dragon with teles and swinging fork, the Super Sports Dragon the same except for Earles forks, and a de luxe version of the latter fitted with a four-speed gearbox and rectified lighting. The 9E engine was listed as an option along with full width hubs and various other parts. With the 200 cc machines came a second competition model for scrambles fitted with the 7E engine, the close ratio four-speed gearbox, Earles forks and swinging fork rear suspension. The Dragon Major also continued.

It was an extensive programme for a small company and lack of capital, shortage of supplies, a railway strike and the lack of suitably skilled labour in the area all combined to bring production to a halt in 1956.

Colours

1954: Dragon—chrome plated tank, mudguards, wheel rims; painted parts in maroon and silver; alloy parts polished.
1955/56: All road—maroon, beige or jade green to order. Competition—silver and chrome.

James

The James story runs very close to that of Francis-Barnett in the postwar years for both came to be part of the AMC group. The James amalgamation occurred in 1951 and from then on the two factories moved closer in their ranges until 1962 when the Francis-Barnett production line was transferred to the James factory at Greet. After that the combined companies drifted into badge engineering.

The company was founded in 1880 by Harry James to build pennyfarthing bicycles and he retired from the business in 1897 when it went public, leaving his manager, Charles Hyde, as managing director. In turn he left and Fred Kimberley became virtual works manager in 1902 when the first James motorcycles were built. He remained with the company, and in 1952 was President of the Manufacturers' Union having served half a century with James.

While he was responsible for the first James machines he had found wooden cylinder patterns in an old shed which indicated some earlier ideas for an engine. The company moved to Greet in 1908 and in the same year introduced its 'safety' model with low frame built entirely with straight tubes, stub axles for the wheels, drum brakes and a four-stroke engine with the inlet valve concentric with the exhaust. Other more conventional models followed including a vee-twin and in 1921 some experiments were carried out with an autocycle. During the 1930s trading conditions worsened and the firm turned to lightweights and Villiers engines.

It was one of these that was built in large numbers for the airborne forces during the war, some 6000 machines being produced. The firm made other war equipment but was hampered by loss of the factory in 1940. It was quickly rebuilt but most of the firm's archives were lost.

After the war the firm picked up the threads by continuing production of the airborne machine and adding an autocycle. The first was re-typed the model ML but aside from lengthened mudguards, toolbox shape, rear carrier and finish, was the same as that supplied to the army. In all over 26,000 were built including the military version.

The machine had a 122 cc Villiers 9D engine mounted in a loop frame with bolted-on chainstays and fitted with blade girder forks controlled by a single central spring. Wheels carried 2.75 × 19 in. Dunlop Universal tyres and the brakes were 4 in. front and 5 in. rear drums. The engine had twin exhausts feeding a transverse silencer mounted in front of the crankcase and connected to a pipe and tubular silencer on the left.

The machine was fitted with a saddle with a cylindrical toolbox below it on the upper left chainstay, while the gearbox was controlled by a lever working in a gate attached to the fuel tank. Rear carrier, bulb horn and direct lighting completed the specification.

The autocycle was driven by the Junior de Luxe engine in a rigid frame with tubular girder forks. Extensive side panels enclosed the engine in the normal style of the type and fitted to the fuel tank. A toolbox was attached to the left of the machine and equipment included saddle, bulb horn, rear carrier and direct lighting.

These two models were continued with few detail changes into 1948 when the first of the Comet machines appeared, a model name that was to continue until 1964. All had 98 cc engines and the first in the line used the 1F unit in a simple rigid loop frame fitted with tubular girder forks. Two models were listed, standard with direct lighting, bulb horn and a little less plating, and the de luxe with battery and rectified lighting system.

With the arrival of the Comet came a very similar model using the 122 cc 10D engine in slightly heavier cycle parts. Again a standard and a de luxe model were offered, while the ML with its 9D unit was dropped. To round off the range for 1949 there was a de luxe machine fitted with the 197 cc 6E engine.

During 1949 122 and 197 cc competition models were introduced, but on a very limited production scale, while in March came news of telescopic forks for export models only. These had been developed by James with the Dunlop Rubber Company and used rubber bushes to act as the suspension medium and to give it a rising rate.

For 1950 the new telescopic forks were fitted as standard to the 122 and 197 cc models, the second of which also became available with plunger rear suspension. They were then given the model names of Cadet and Captain to be used by those capacities from then on. At the same time the old autocycle was replaced by a new model using the 2F engine in a frame with tank and engine shields to suit it. Girder forks were retained.

In February 1950 a further 98 cc model, later to be called the Commodore, was seen in prototype form, this being a Comet fitted with extensive weather protection. This comprised enclosure panels that ran from legshields at the front back to the rear wheel spindle, and from footrest level up to the cylinder head. A

James 122 cc ML of 1948 with 9D engine and based on wartime model

Standard James Comet of 1949 with 98 cc 1F Villiers engine

windscreen was also fitted.

It joined the range for 1951 which was otherwise continued with detail changes only and this policy was maintained into 1952.

There were some changes for 1953 with the Comet and Commodore models fitting the 4F engine unit, while the autocycle remained unaltered. The standard Cadet was replaced by a utility model numbered J5 and fitted with the 13D engine. It had a new frame with plunger rear suspension and simple telescopic front forks using springs as the suspension medium. The deluxe Cadet continued, also with plunger suspension, as did the Captain, and a dualseat was offered as an option. New for the Commando trials model were hydraulically damped front forks fitted with springs and hydraulic end stops.

In June 1953 a telescopic fork conversion became available for all 98 cc models made from

James Superlux autocycle of 1949, 2F engine replaced earlier Junior De Luxe

1950 Cadet de Luxe with 122 cc Villiers 10D engine in rigid frame. Note wing emblem

1952 Cadet with plunger frame, usual Villiers emblem on chaincase

1948 onwards and was a standard fitment on the Comet for 1954.

That year saw the autocycle still in the lists but only up to July, while a single Comet, the J11, continued. It was further modified by the addition of plunger rear suspension as well as the telescopic forks and continued the tradition of offering basic transport. The enclosed Commodore was dropped as was the older Cadet J6, but the J5 version continued.

In the 200 class the Captain with 6E engine was replaced by the K7 powered by the 8E fitted in a frame with swinging fork rear suspension and three-rate spring telescopics at the front. Both front forks and rear units were made by James. Neat boxes were fitted under the nose of the dualseat on each side of the machine, one for tools and the other housing the battery. The same frame was also used for the new Colonel model K12 which was fitted with the 1H engine, and both machines had well valanced mudguards.

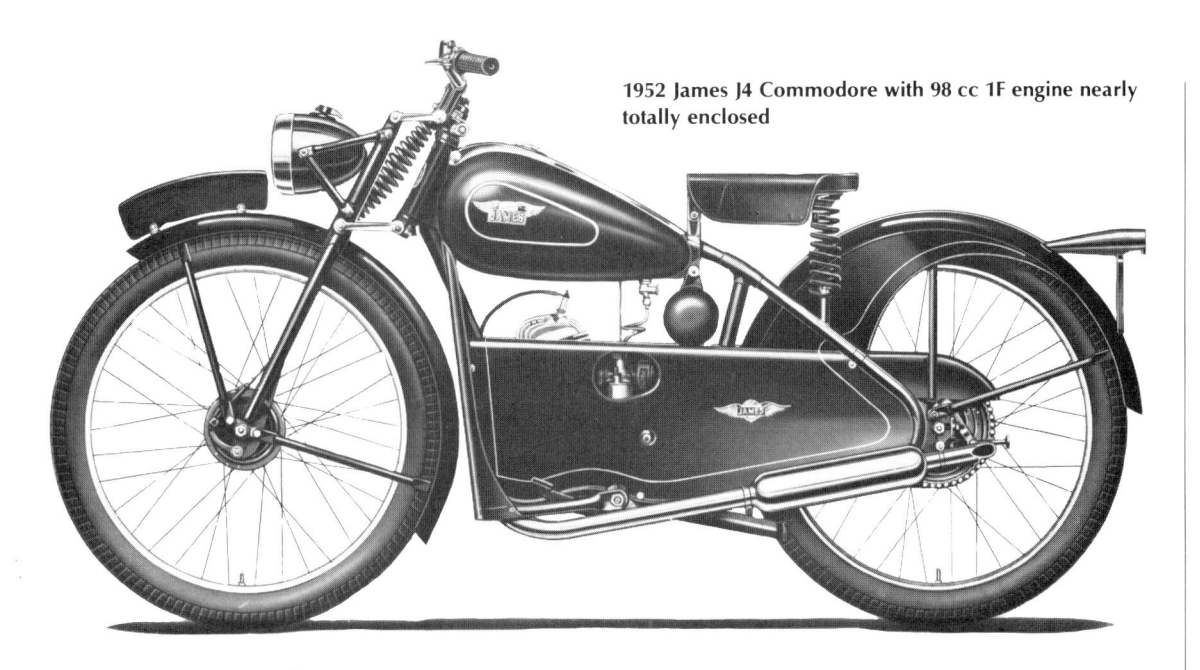

1952 James J4 Commodore with 98 cc 1F engine nearly totally enclosed

Competition James, the 1953 Commando J9 with 7E engine and four-speed gearbox

The 1955 Colonel K12 model with 1H engine carrying James badge

On the competition front the Commando model J9 was fitted with the 7E engine in its rigid frame and was joined by the Cotswold model K7C. This was aimed at the scrambles market so had the close gearbox ratios and a swinging fork frame. Curiously it retained a silencer on a low level exhaust pipe, a centre stand and a toolbox but did have a seat, unlike the trials J9 which kept to a saddle.

1955 brought consolidation to the range with the 98, 197 and 225 cc models continuing with only a few changes. The Cadet J5 was replaced by the J15 but the only real change was that the larger 30C engine was fitted. There were alterations to the handlebar fixings and the headlamp shell was lengthened to take speedometer, lighting switch and ammeter. Also, full width hubs were adopted on the road models. The 197 and 225 cc machines were all fitted with new front forks with hydraulic damping, the latter feature previously only found on the competition models.

There were a number of machine changes for 1956 with the Comet and Cadet both being replaced by new models in a common frame. This was built up from tubes and pressings with the rear swinging fork controlled by a pair of forward mounted springs which were hidden by the centre enclosure to which the deeply valanced mudguard was attached. The Cadet was fitted with new telescopic forks with tension springs and compression oil damping and both models had an alloy expansion chamber mounted beneath the gearbox with twin tailpipes. A single seat went on top of the rear enclosure panelling and a pillion section and footrests could be added to the Cadet.

The road Captain continued with a revised silencer and modified front forks, while the Colonel was also equipped with the new exhaust system. The competition K7C carried on, but the J9 was replaced by the K7T which, like the scrambler, had a swinging fork frame. All models had a plastic push-in tank cap fitted.

In 1957 the Comet was made available with the footchange 6F as well as the 4F and an electric horn was fitted as standard. A useful addition was a one-piece optional leg screen that ran under the engine a little and had a centre hole fitted with a perforated panel to let cooling air through to the motor. With both the new and old engines came a conventional exhaust system in place of the earlier alloy box. The Cadet was fitted with a horn and received in addition a new silencer based on those used for larger models, and also lowered gearing. The rear tyre on the Captain went up a little in section and it and the Colonel were fitted with mudguards with deeper valances, an improved chainguard and a stiffer fork crown. The four speed gearbox option for the 197 was no longer available. The Commando was fitted with a rock shield to protect the flywheel magneto, a revised run for the exhaust system and a rear chain guide.

This range was joined by a new model, the L25 Commodore, thus reviving the enclosed Comet name. The new machine was very different, being powered by the 249 cc AMC engine also used by Francis-Barnett. It went into a frame built up with a single top and down tube which bolted

Left **Line of James Captain K7 machines for the Forestry Commission**

Right **Trials model K7T in 1956 with 7E engine and low exhaust**

to a centre section and rear mudguard built up from pressings to give a large degree of enclosure. This was supplemented by a deeply valanced front mudguard fitted to the telescopic forks, and a full case for the rear chain. The rear units were Girling but most of the main parts of the machine came from the AMC group. A dualseat and full equipment completed the bike.

The range was contracted a little for 1958 with the two competition models and the Colonel going, while the Comet and Cadet continued with the Captain which was fitted with the 10E engine and Girling rear units. New was the L17 Cavalier fitted with the 171 cc AMC engine in the Cadet cycle parts but with a 5 in. front brake and the hydraulic damped front forks.

During 1958 the Commando trials model reappeared briefly fitted with the 10E engine and equipment to suit its use, including upswept silencer, alloy mudguards, no horn, and saddle. It was soon superseded by the L25T, a similar machine powered by the 249 cc AMC engine and partnered by the L25S with the same engine but developed for scrambles.

The two competition machines shared a common frame with single main loop and swinging fork rear suspension controlled by Girling units. The telescopic front forks were different with the trials model being fitted with James legs with the hydraulic damping stiffened, while the scrambler used Matchless Teledraulic forks. Engines and gearing were modified to suit the machine's use, as were the cycle parts.

The five road models were continued with few changes, although the Cadet became fitted with a dualseat as standard. The swing to AMC engines continued for 1960 when the L15 Cadet and K7 Captain were replaced by the L15A Flying Cadet and L20 Captain with 149 and 199 cc AMC power units. The L17 Cavalier was dropped but the Comet and the three L25 machines all continued.

The Flying Cadet followed the lines of the earlier model for its cycle parts which were very similar to the Comet, while the new Captain was styled along the lines of the Commodore but retained a tubular frame and rear fork. Thus only the Comet used a Villiers engine and this range

103

James trials K7T model in 1957 with proper exhaust system

1960 model L25 Commodore with 249 cc AMC engine

was continued on into 1961 with only colour changes.

During that year a Sports Captain was added to the range, this having improved engine performance, rearsets, rear-facing gear pedal, folding kickstart lever and right footrest, 6 in. front brake, flyscreen, dropped bars and alloy mudguards.

For 1962 it was joined by another new model, the M25 Superswift, based on the L25 but fitted with the Villiers 2T engine. The other road models continued with little change, although the option of the 4F engine for the Comet was discontinued. The L25T Commando was revised with a shorter wheelbase and an offset right rear damper unit which allowed the silencer to fit between it and the rear tyre. The tank was moved back to increase the steering lock and the front brake went down to 5 in. while the seat, footrests and pedals were all altered a little.

During 1962 the appearance of a James scrambler fitted with an A series Villiers engine with Parkinson conversion gave a clue to the 1963 models. When these were announced the three L25 machines had gone and in place of the competition versions were the M25T for trials fitted with the Villiers 32A engine, and the M25R for scrambles with the 36A and the Parkinson conversion. Both machines were fitted with Norton forks in a similar manner to their Francis-Barnett counterparts, the latter having moved their production line into the James factory during 1962.

The amalgamation also showed up in the adoption of Arden green for the Comet and the replacement of the L15A by the M15 which was common to the Francis-Barnett model 95. Thus it repeated the tubular spine frame and short humpy tank. The two Captain models and the twin all continued and were joined by a Sports Superswift. This used a tuned version of the 2T engine in a tubular frame with the same features as the Sports Captain and both were fitted with very nice Italian style petrol tanks.

1964 saw the Comet, Cadet and both Captains continue but the Superswift was dropped while the sports twin had the 4T engine fitted. Both it and the Sports Captain changed their front tyres to 2·75 × 19 in. to improve handling. The two competition models also continued but early in the year the scrambler M25R was replaced by the M25RS powered by the Starmaker engine. Aside from the engine and a 6 in. rear brake, the machine was as its predecessor.

The industry was falling on hard and difficult times so James left their seven models alone when they announced the 1965 range but one, the long running Comet was dropped before the new year came in. During 1965 the M15 was replaced by the M16 which retained the 149 cc AMC engine, assembled by Villiers at Wolverhampton, in a very conventional set of cycle parts.

Thus the four road models and two competition ones continued on into 1966, but in October of that year production of James motorcycles ceased.

Colours

1946/47: **Superlux** autocycle—tank in black with Argenize (a special silver sheen finish developed by James) lining; black frame, forks and mudguards; black wheel rims lined red; Argenize hubs and side shields which were lined blue **122ML**: maroon tank with Argenize side panels; Argenize wheel rims and hubs; chrome plated front exhaust pipes, black silencers and rear exhaust pipe.

1948: **Superlux**—Argenize silver fuel tank, maroon engine shields. **122ML**—maroon with light blue tank panels lined gold; maroon hubs; Argenize silver wheel rims, exhaust system as 1946.

1949: **Superlux**—as 1948. **98**, **122** and **197** cc models—maroon with light blue tank panels lined gold; maroon hubs; Argenized silver wheel rims. De luxe had motif on leading edge of front mudguard, silencer and strip on tank top chrome plated; standard was minus motif and not plated.

1950/52: All models—maroon with light blue, gold lined tank panels as 1949.

1953: All models—maroon with lined tanks;

Above **James Cadet M15 of 1962–65 with 149 cc AMC/Villiers engine**

Top **James L20 Captain, 199 cc AMC engine, built 1959–62**

Centre **1960–62 model L1 Comet with 98 cc Villiers 6F engine and semi-enclosure**

Villiers Singles & Twins

105

James M25 Superswift twin from 1962 with 2T engine and some enclosure

Superswift M25 in 1963, full rear chaincase and white-wall tyres

Argenized silver wheel rims; maroon hubs; chrome plated exhaust systems.
1954: All models—as 1953, alloy mudguards for competition models; lining in silver.
1955: All road—as 1953, hubs polished. **J9**, **K7**—as 1954.
1956: All models—as 1955 with brake back plate in Argenized silver finish. During year **K12** option added; Seagull grey with pastel grey tank panels lined in black and silver; grey centres to chrome plated wheel rims.
1957: All models—Martial dark grey with gold lined royal blue tank panels; seat in black. Option for **147**, **197** and **225** road models—maroon with gold lined pastel grey tank panels; seat in pastel grey wheel rims chrome plated with centres in machine colour, gold lined; road models have name transfers on centre section sides. **L25**—as other road models.
1958: All models—including **L17**—as 1957.
1959: Road—as 1958 except Comet in maroon and grey with chrome plated bars. Competition—tank red with silver side panels; black frame and forks; polished alloy mudguards; chrome plated wheel rims.
1960: **L1**, **L15A**—maroon with tank top part in grey with gold line; grey seat; chrome plated wheel rims, bars, headlamp rim and exhaust system; name transfer on side of centre panel. **L20**, **L25**—black frame, forks, rear fork, rear units, centre stand on **L20**, chainguard on **L20**, with mudguards, headlamp shell, tank, centre panels, centre stand on **L25**, chainguard on **L25**, in Caribbean blue for **L20** and Stromboli red for **L25**; chrome plated wheel rims and tank side panels; White-wall tyres on **L25**. **L25T**, **L25S**—as 1959.
1961: **L1**, **L15A**—as 1960 in Stromboli red and grey. **L20**, **L25**—as 1960. **L20S**—Caribbean blue frame, forks, tank; silver chainguard, rear number plate mounting; chrome plated wheel rims, tank side panels, tool and battery box lids; polished alloy mudguards **M25**—tank, forks, frame, chaincase, lower part of tool and battery box lids, lower part of rear enclosure and rear mudguard in flamboyant Riviera blue; front mudguard, upper lids and rear mudguard in silver sheen; gold lining between colours and to outline front number plate area on front mudguard; chrome plated tank panels and wheel rims. **L25T**, **L25S**—as 1959.
1962: As 1961.
1963: **L1**—Arden green, white tank flash with

1964 Sports Superswift M25S with revised controls, flyscreen and 4T motor

1965 James model M16 with 149 cc AMC engine built by Villiers. At Blackpool Show

gold lining; chrome plated wheel rims. **M15**—Stromboli red with silver tank panels; chrome plated wheel rims. **L20**—Caribbean blue; chrome plated tank side panels and wheel rims. **L20S**—Caribbean blue; silver tank panels lined gold; alloy mudguards; chrome plated side lids and wheel rims; silver chainguard and rear number plate support. **M25**—as 1961. **M25S**—as **L20S** in Riviera blue and silver. **M25T**, **M25R**—as **L25T** in 1959.

1964/65: All models—as 1963.

1965: **M16**—metallic green tank, mudguards, side panels; black frame, forks, headlamp shell, chainguard; chrome plated wheel rims.

1966: All models—as 1965.

Mercury

This shortlived company, based in Dudley, was mainly involved with small capacity scooters but did produce a motorcycle for the 1957 season. This was called the Grey Streak and powered by the two speed 6F engine mounted in a frame with telescopics and swinging fork rear suspension. Small full width hubs were fitted along with a 2 gallon tank, direct lighting and a dualseat.

Plans for the company were perhaps rather grandiose for its size, as indicated when a 98 cc scooter was announced in January 1958 and said to be tooled for an output of 15,000 units that year.

Planning is one thing, optimism another and little more was heard of the firm.

New Hudson

This make faded from the motorcycle scene in the 1930s when the firm became the sole manufacturer of Girling brakes. After the war the name was revived as a member of the BSA group and the autocycle the company built continued as a standard for the type until the moped took over its job of providing minimal transport.

The 1946 model was propelled by a Villiers Junior de Luxe engine and was a fairly basic device with heavy duty, ladies-style bicycle frame from which the engine was hung. The fuel tank tucked between the frame tubes to give the typical autocycle line.

1953 example of the New Hudson autocycle, a model that typified the breed

In 1948 pressed steel blade forks were added along with engine shields so that with the extensive chainguards on both sides, the works were well enclosed. Slots and holes gave access to the choke and petrol tap but the handlebar area reflected the machine type with a clutter of controls, cables, bulb horn and headlamp. A rear carrier went behind the saddle with the number plate attached to it.

In the middle of 1949 a revised machine was announced fitted with the Villiers 2F engine. This entailed a new loop frame to run under the power unit and a revision to the shape of the petrol tank and engine shields. That aside, the specification was as before with blade girders, small drum brakes and no rear suspension. Direct lighting was powered from the flywheel magneto, while a dry battery looked after the parking light need.

The machine remained virtually unchanged for several years, although the centre stand was strengthened for 1952 and, during that year, the colour was changed to green.

It was not until 1956 that any further change occurred when the styling was extensively altered, although the engine remained the faithful 2F. A new frame carried it and was fitted with tubular girder forks of improved appearance. The fuel tank and engine shields were still blended together and went on into the chaincase to produce one continuous form. Legshields were available as an extra and both mudguards had partial deep valances to improve rider protection and style.

The finish was brightened up with more chrome plating and a rear carrier, speedometer and windscreen also available as extras. Unfortunately these and the new styling were no real contest to the many mopeds coming into the country, many of which performed as well and were cheaper to tax and insure. So the day of the autocycle came to an end and with it went the New Hudson make from motorcycle lists.

Colours

1946/49: Black with cream lining, cream panel on engine shields, name on tank and symbol on chainguard. Chrome plated headlamp rim, bars, minor controls.

1949/52: Black, red tank panels, gold lined. Transfer in gold on tank panel and engine shields, the latter gold lined. Chrome plating as 1946/49.

1952/56: Dark green, cream tank panels, gold lining and transfers plus plating as in 1949.

1956/58: Maroon finish, gold lined; chrome plated wheel rims and other items as before.

Norman

Norman re-entered the motorcycle market after the war with the same range they had offered before it, an autocycle and a 125 cc motorcycle. Both were typical of the austere period with the first equipped with the Junior de Luxe engine in a rigid frame and the second with the 9D unit.

These two continued until late 1948 when a three model range was announced for 1949. The autocycle continued but fitted with a 2F engine in a revised frame and called the model C. The power unit was well enclosed by side panels above which went the tank. Tubular style girders were fitted to the rigid frame and equipment included saddle, rear carrier, bulb horn and direct lighting.

The autocycle was joined by the B1 and B2 motorcycles fitted with the 10D and 6E engines. The cycle parts were the same in both cases with rigid frames, light telescopic forks with oil damping, 5 in. brakes and 2·75 gallon tank. The standard models had direct lighting, while the de luxe came with rectifier, battery, electric horn, ammeter and rear stop lamp. The exhaust pipe had an expansion chamber in it close to the cylinder before running along to the silencer on the left.

For 1950 the range was supplemented by the model D, a 98 cc motorcycle, with the 1F engine in a rigid frame with light tubular girder forks changed in 1951 to simple telescopics. As was usual Norman practice both standard and de luxe versions were available with direct or rectified lighting.

Aside from a reduction in chrome plating in 1952 due to a world shortage of nickel, the range remained unaltered apart from detail improvements. Early that year the B1S and B2S were introduced with swinging fork rear suspension built onto the rear of the existing frame. The fork pivoted on a spindle pressed into a frame lug and its movement was controlled by a pair of laid down spring units which ran from the rear wheel fork ends to a cross spindle in the frame lug connecting top and seat tubes. This arrangement enabled the suspension to be added with the minimum of redesign to the frame but did give the units a considerable length and awkward operating angle. A dualseat was fitted as standard and supported by seat stays bent out and round the spring units.

The B1 was superseded by the model E for 1953, this being an economy job with rigid frame, telescopic forks and available in standard or de luxe form. It was fitted with the 10D engine while the B1S used the 13D. The models C, D and B2 continued with the two new springers, and a competition 197 was added to the range, the B2C. This used the 6E engine in a rigid frame with telescopic forks and was supplied without lights but suitably equipped for trials.

For 1954 there were changes to both engines and frame. The latter affected the B2S which received a more conventional form of swinging fork rear suspension controlled by near vertical Armstrong units reacting against a bolted-on sub frame. The B1S continued with the original Norman arrangement. The engines fitted into the 197 cc models became the new 8E units with the choice of three or four speeds for the B2S but not the B2. The B2C had the same gearbox choice and was equipped with the 7E unit. The model E was dropped and the C unchanged, but the D engine became the 4F unit.

The 1955 range brought further changes with the B1S increasing in size to 147 cc with the 30C engine but retaining the spring frame with the long Norman units. The rigid B2 went but a new and larger model appeared using the 242 cc Anzani engine in the swinging fork frame, this being fitted with Armstrong bottom link forks also used for the B2S and an option on the competition model. The D and C machines received minor mudguard changes.

The firm did fit a 322 cc Anzani engine into one frame but only as an experiment and it was solely

Norman model B1 de luxe in 1949 with early 10D engine without wing emblem

1953 model D de luxe with battery, 98 cc 1F engine

on that basis that *Motor Cycling* were allowed to try it attached to a sidecar. They even had to arrange for the sidecar to be fitted.

There was no model D in 1956 as it was dropped along with the B2C, but that was replaced by a new version, the B2SC fitted with the 9E engine. The frame was new with swinging fork rear suspension and the bottom link forks at the front. These were also fitted to the B1S so all models, except the C, had them. Full width 6 in. hubs went onto the competition model and for 1957 this style became common to the road models, although the diameter remained at 5 in.

That year brought a reappearance of rear enclosure panels that had been tried two years earlier but had failed to go into production. The new ones were well styled and one carried the electric horn. Less noticeable was another engine change for the B1S to the 148 cc 31C unit which became available as an alternative to the 30C. Unlike this, which continued with only three gears, the new unit was supplied with four. During 1957 the autocycle was dropped as the firm had launched a moped with Sachs engine which effectively replaced it.

The 1958 range was little changed but did see the Anzani engined twin go, to be replaced by the B3 powered by the Villiers 2T. This followed the same general lines with the bottom link forks but had the 6 in. brakes in full width hubs and a larger fuel tank. The B2S and B2SC both continued and for 1959 the road model was renamed the Roadster and fitted with the 9E engine unit. This went into the B3 frame and was available as a standard model with three speeds or a de luxe with the option of four. In either case the larger tank was fitted. The two Roadsters were joined by a Sports version with down-turned bars, polished mudguards, a smaller tank and a narrower section front tyre. A similar style B3 twin was also introduced in the same format.

At the end of 1959 the B1S with either the 30C or 31C engine was dropped, but the other machines continued for 1960 until October. Then the B2SC was replaced by a new trials model, the B4C with 9E engine, and this was joined by two 250 cc competition models with the same designation. All had the same cycle parts with bottom link forks and swinging fork rear suspension, while engines were the 32A for trials and 34A for scrambles. The two twins received a new frame and larger fuel tank to be recoded B4 in Roadster and Sports styles. The Sports was fitted with low bars, a perspex screen and quick action filler cap, while the Roadster had deeply valanced mudguards and rear enclosure panels that ran back to the rear number plate. The B2S continued also in Sports and Roadster style and with standard or de luxe specification.

The motorcycling days of the company were nearing their end for late in 1961 all the B2S variants and the B4C in both engine sizes were dropped to just leave the B4 twin models. In 1962 these too went to bring Norman motorcycles, along with their mopeds, to a close.

Colours
1949: **C**—maroon. **B2**—black.
1950: **C**—maroon. **B1**, **D**—maroon and chrome tank. **B2**—black frame and forks; tank chrome plated, panelled red, name in white and gold.
1951: **C** and **D**—black with maroon tank, chrome plated exhaust system, bars, controls, wheel rims, headlamp rim. **B1** and **B2**—black with tank as 1950 **B2**; chrome plating as **C**.
1952: All models—reduced chrome plating, tanks in maroon with gold lining, wheel rims in silver.
1953: All models—black frame and cycle parts, red tank with gold lining, chrome plated exhaust system, bars, controls, wheel rims, headlamp rim. Options—mid-green or old bronze with two-tone silver.
1954/55: All motorcycles—as 1953, option of mid-green. **C**—deep red.
1956: Road models—green lustre with silver lining.
1957: All models-metallic dark blue frame, forks, tank, headlamp shell, rear enclosure main part; light blue tank top panel, centre panel of rear enclosure, mudguards.
1958: **B3**—black, blue option.
1960: All models—maroon and black.
1961: **B4** Sports—red tank with gold lined ivory panels, red frame, forks, cycle parts; alloy mudguards. **B4** Roadster—Cypress green lined grey.

Above **Norman model TS built from 1955–57 with 242 cc Anzani engine**

Top **1953 Norman competition model B2C with 197 cc engine**

Centre **1954 B1S de luxe with 13D engine and Norman rear suspension units**

111

Above **1957 Norman TS de luxe with Anzani twin power at showtime**

Above **B3 Sports twin from 1959–60 with lowered bars and flyscreen**

Below **Norman B3 twin with 2T engine on the road in 1958**

OEC Atlanta ST2 in 1953 with 6E engine in sprung frame

OEC

The OEC was one of several makes that was revived by the ubiquitous Villiers engine after the war. In pre-war days it was known for record-breaking, duplex steering and a swinging fork rear suspension design that was linked to plunger boxes as far back as 1928.

The machines came from Portsmouth and were re-introduced in 1949 with the model name Atlanta, which was used for all fitted with two-stroke engines from then on. Two models were announced, both fitted with three-speed Villiers engines, one of 122 cc and the other of 197 cc. The frame was a simple loop with the rigid rear part bolted to it. OEC telescopics were fitted containing both load and rebound springs and with gaiters to keep dirt at bay. Drum brakes were used, the front smaller than the rear, and a substantial rear carrier and lifting handle fitted behind the saddle.

Rectified lighting was fitted so the machines were equipped with battery, electric horn and stop light as standard. These models were given the codes of S1 and S2 for the two sizes and were joined in 1950 by the D1 and D2 which had direct lighting but otherwise were to the same specification. With them came two competition models, the C1 and C2 with upswept exhausts, competition tyres, direct lighting and alloy mudguards.

At the end of the year the range was expanded further for 1951 by the appearance of an optional spring frame for the four road machines. This was of the swinging fork type with vertically mounted units each containing a spring screwed to the mounting lugs so able to work in compression and tension. The whole suspension assembly was self-contained so could be bolted to any existing postwar OEC machine with ease. The front end also received some attention with a number of detail changes being made to the fork crowns, which became malleable castings.

113

OEC model S1 Atlanta in 1949 with 10D engine, battery lighting and rigid frame

In addition to the standard range of machines, OEC would also supply models suitable for grass track or speedway use, even listing a light single-seater sidecar.

For 1952 all the two-strokes continued unchanged and were joined by the Apollo, a model with a 250 cc sidevalve Brockhouse engine in a rigid frame built from square section tubing. For the next year the C1 was dropped but all the road models with 122 or 197 cc engines, direct or rectified lighting and rigid or pivoted fork suspension continued as before.

Also new for 1953 was a new swinging fork frame built from square section tubing and used for a second version of the Apollo and two 197 cc models, one for road use and the other competition. For these frames Girling rear units were fitted along with revised OEC front forks.

The most novel feature of the range was the transmission of the new competition model, the ST3. To ensure that the rear chain tension was kept constant a countershaft was run on taper roller bearings mounted on the swinging fork itself to be concentric with the pivot. A sprocket was cut on each end so that the final drive was by a very short chain on the left back to the countershaft and then by another chain on the right to the rear wheel. The middle chain was completely enclosed and the final one had the usual guard.

During 1953 the range contracted with all the early models being dropped, while the newer 197 cc ones continued. They were joined by two of 122 cc, one in a rigid frame and the other with swinging fork rear suspension.

All four continued with the Apollo and the sidecar into 1954 and the two fitted with 197 cc engines had these changed to the 8E unit for the road machine and the 7E for the competition. Not too many of these came to be built, however, as the firm was falling on difficult times and late in 1954 production ceased.

Colours

1949: **S1** and **S2**—black frame and forks; polychromatic silver mudguards, chainguard, toolbox and fuel tank which also had angled panels in blue with lettering in gold. Chrome plated wheel rims.
1950: **S1** and **S2**—as 1949. **D1** and **D2**—as **S1**. **C1** and **C2**—as **S1** except for alloy mudguards.
1951/52: All models—as 1950.
1953/54: All models—as 1950 except tank blue lined, not panelled and wheel rims in silver. Handlebars painted black.

Panther

The Panther machines built by Phelon and Moore from Edwardian days are best remembered for their big singles where the engine was used as the frame down tube. Along with this type they built a variety of others such as the transverse vee twin Panthette of 250 cc and the similar capacity Red Panther sold from Pride & Clarke in South London at under £30 in the 1930s. There was even a two stroke Panther in those days fitted with a Villiers engine.

After the war production concentrated on the 600 cc single and a pair of smaller four-strokes with vertical cylinders sold in touring or trials forms. This range was joined for 1956 by a pair of two-stroke lightweights using Villiers 197 cc engines.

The first of these, the 10/3, was fitted with a three-speed 8E unit, while the second—the 10/4—had the 9E and four gears. The use of two different engines entailed two versions of the frame to suit the mountings and both were fully welded with swinging rear fork pivoted on Silentbloc bushes and controlled by Armstrong units. At the front went Earles forks controlled by Girlings and the fork design was of a very sleek appearance and not easy to distinguish from telescopics.

Full width hubs were used for both wheels which had 3·25 × 18 in. tyres and 5 in. brakes. Mudguards were deeply valanced and the rear one merged into a central enclosure panel below the dualseat. It made for a smart machine and a well equipped one as the enclosure housed the battery and rectifier of the electrical system along with the horn and tools.

For 1957 the range of single cylinder two-strokes was augmented with the 10/3A which was the 10/4 but fitted with the three-speed 9E unit. In addition a twin was added as the model 35 using the 2T engine in the same cycle parts. All models had both brake diameters increased to 6 in. but otherwise remained unchanged. A further model using the 2H engine and called the 25 was proposed but failed to go into production.

The four models were joined by the 35 Sports for 1958, this being a twin fitted with a tuned engine and a 7 in. front brake. It was given a different finish to the standard models. It was supplemented late in the year by the 45 Sports on similar lines but fitted with the 324 cc 3T engine unit. This in turn was joined early in 1959 by the model 50 Grand Sports with a tuned version of the 3T, telescopic front forks and 8 in. brakes on both wheels. On this model the exhaust pipes were siamezed at first but this was soon changed to the normal Villiers twin exhaust system, while the rear wheel was partially enclosed by extending the centre panelling back to the rear number plate.

The two stroke range thus stood at seven models for 1960 and the 35 and 45 joined the 50 in the fitting of Panther telescopic front forks, while the 35 Sports also gained the rear enclosure used by the 50.

During 1960 the range contracted a little with the 10/3 and its 8E engine being dropped along with the standard model 35. The other five ran on into 1962 when a further contraction took place. Late in the summer the two 197s went along with the model 50 and a month later the 35 Sports.

This just left the 45 Sports and this was joined by the 35 which reappeared after nearly two years absence with telescopic forks, the rear enclosure and the 7 in. front brake. The 45 also had the rear enclosure fitted.

These two models continued until 1964 when the 45 went but the 35 ran on, with the big 650 cc single for a few further years. Late in 1965 it gained electric starting and in this form continued until the company ceased motorcycle manufacture late in 1968.

Colours

1956: **10/3** and **10/4**—maroon with gold tank lining; round plastic tank badge; chrome plated

Part Two Machines

Above **Panther model 50 with 3T engine, telescopic forks and 8 in. brakes**

Above **1956 model 10/4 with four-speed 9E engine and very sleek Earles forks**

Below **Panther 35 from 1957–59 period with 2T engine and model 10 styling**

wheel rims, headlamp rim, exhaust system, bars, controls.
1957: **10/3A** and **35**—as **10/3**. **10/3** and **10/4**—as 1956.
1958: All above models—as 1957. **35 Sports**—pearl grey, red and gold trim.
1959: **50**—red frame, forks, mudguard, rear enclosure, tank; chrome plated tank panels.
1960/61: All **197** models—red. **35 Sports**—sea mist grey, chrome plated tank panels. **50**—as 1959.
1962: **35 Sports**—as 1960. **45** and **50**—devil red, chrome plated tank panels.
1963: **35** and **45**—as 1962 '**45**'.

Radco

The roots of this make, built by the Birmingham firm of E. A. Radnall, go back to just before the first world war. They built a variety of machines in the twenties but were one of many that ceased motorcycle production in the early thirties. To keep going they turned to the making of parts and accessories.

In 1954 they reappeared on the motorcycle scene with the announcement of the 98 cc Ace. This was a simple utility mount powered by a two speed 4F engine which was housed in a straight-forward single loop frame with rigid rear end. The front forks were Metal Profile bottom or leading link type and small drum brakes were fitted.

Equipment comprised direct lighting, bulb horn, cylindrical toolbox, saddle and deeply valanced mudguards. Finish was black with matt silver wheel rims and metallic silver tank which carried the original style Radco transfer and had a contrasting red centre stripe.

Following the announcement in January the machine was listed in a buyer's guide in March but failed to reach production and no more was heard of it.

The firm did continue, however, and in 1966 exhibited a 75 cc runabout at Earls Court. The machine was very basic with a side valve, single speed Villiers engine started with a pull cord. No suspension was fitted, the machine relying on its small doughnut tyres to absorb road shocks. It had minimal controls, no lights and was called the Radcomuter, being intended to carry its owner to his car and then ride in the boot.

Nothing more was heard of that machine either and the company reverted to other products.

Rainbow

The Rainbow was a prototype machine built by a gentleman of that name in 1950 and, while it could have been considered an autocycle, it was really a form of step-through. The power unit was the readily available 1F, two-speed Villiers built into a machine with many unusual but practical ideas.

The aim of the builder was to create a link between bicycle and motorcycle and to do this he aimed for an enclosed engine, low centre of gravity, built-in legshields and ease of mounting for the rider.

To achieve these aims the frame followed the layout to be adopted years later for step-throughs but modified to suit the Villiers mountings and the builder's facilities. It comprised twin large diameter tubes which ran down from the steering head under the engine unit and then up and back to the rear wheel spindle. Cross tubes were welded in to join them and rising tubes supported the Dunlopillo seat mounted on a cantilever bracket and adjustable for tilt. It was a good deal different from the conventional bicycle saddle then much in vogue.

At the front went single blade, tubular girder forks with rubber band suspension, while the rear end was rigid. The wheels were completely different for the front was a 26 in. diameter heavy duty bicycle type, while at the rear went a

Part Two Machines

The Rainbow built in 1950 to demonstrate the designer's ideas

2·50 × 20 in. tyre to help keep the weight and saddle height low. Both wheels had small drum brakes.

Engine enclosure was by means of a domed steel cover which sprang over rubber blocks to conceal the top of the engine and had the appearance of a petrol tank. In fact the petrol was carried in a transversely mounted cylindrical tank clipped behind the steering head and whose ends became kneegrips with the rider seated.

On either side of the machine, below the engine shield was a circular shape and, while that on the right was the usual flywheel magneto, it was balanced on the left by the silencer. This was round and a spring blade with rubber buffer rose from the left footrest to touch it. This item did two jobs, the obvious one of keeping the rider's foot clear, but also it deadened the exhaust sound. The silencer allowed the gases to swirl within it and then to leave by a pipe from its centre.

The machine was fitted with legshields and had many neat detail points in its contruction. Like most driven by the 1F engine, it could reach just over 40 mph and climb steep hills in its lower gear. It did this in a style not to be seen again until the C50 Honda became popular.

Rather sadly the Rainbow failed to go into production so only the prototype was built to show the innovative flair of its designer.

Raynal

This was a pre-war make of autocycle that was reintroduced to the public in 1946. The machine, called the Popular, was fitted with the Villiers Junior de Luxe engine in a typical rigid autocycle frame with light girder forks.

During the year the Popular was replaced by the de luxe which had a clutch and direct lighting. Tyres were 26 × 2 in. and the machine conventional for the times.

It continued unchanged up to late 1950, when its production ceased.

Sapphire

Roger Kyffin built Sapphire machines for moto-cross or trials using Villiers and Triumph engine units.

Construction was conventional for the 1960s and the machines were nicely made with good detail work. As with many such makes, the numbers built were few but the owners could count on the builder's personal attention to needs and often the result was very satisfying to both parties.

Naturally the make was dependent on its supply of engines and so in time had to give up the Villiers units.

Sapphire built by Roger Kyffin using Starmaker engine

Raynal autocycle in 1939 with Junior engine, JDL used postwar

Scorpion

This company was launched early in 1963 to market a trials machine in kit form or fully assembled. It used the 9E engine as standard but would also accept larger units including the BSA B40. The Villiers ones were offered as they came from Wolverhampton, or with the Marcelle or the Parkinson coversions.

Much of this flexibility came from the frame engine loop which bolted into position and could readily be changed for another to suit the engine fitted. As well as the 9E the frame would accept either the 32A or the Starmaker to trials trim. The main frame comprised three large box sections made from folded up sheet steel and welded together. They formed the top member which ran back from the steering head, the seat beam to which it was joined and which ran down behind the engine to support the rear fork and the footrests, and a rear member with two arms that ran back to support the tops of the rear suspension units.

The engine loop was formed in square section tubing and this was also used for the rear fork. At the front went leading link forks controlled by Armstrong units, as used at the rear. Wheels were wire spoked onto Motoloy hubs with 6 in.

Scorpion 1965 leaflet showing their road race machine with spine frame

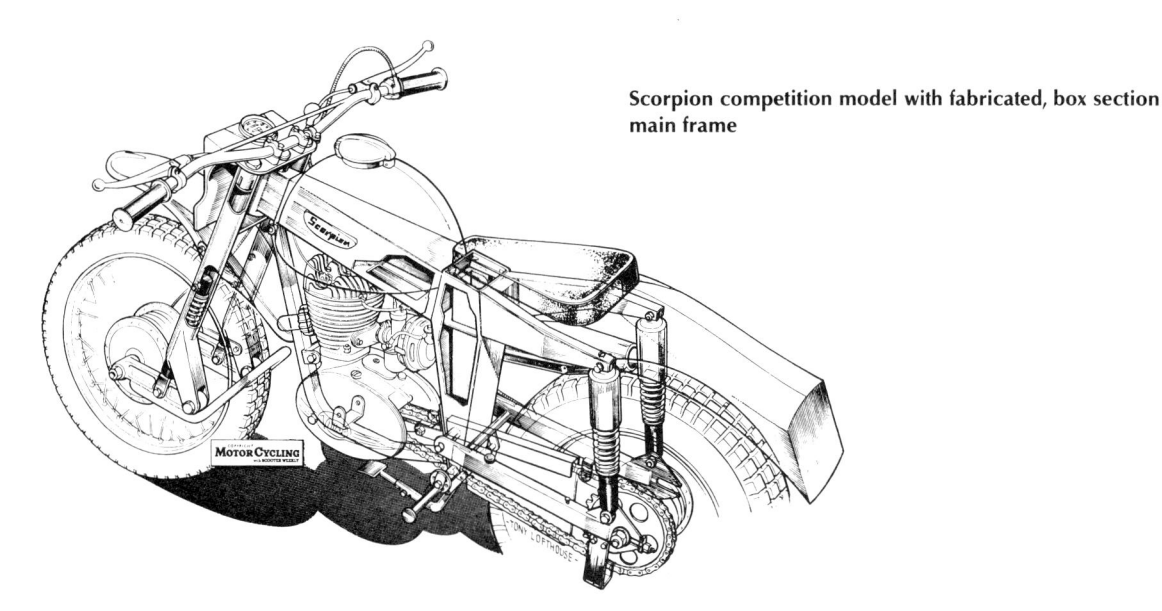

Scorpion competition model with fabricated, box section main frame

brakes and conventional 21 in. and 18 in. competition tyres were fitted.

Fibreglass was used for the tank, of either one or two gallon capacity, the front mudguard and the rear tail unit which ran foward to act as a seat pan and back to the rear number plate. It was also used for a tool container that was carried in the vertical beam under the seat. Equipment included speedometer, bulb horn, braced bars and competition plate, while as standard the machine was fitted with controls and self-locking nuts in aluminium, steel ones being an option.

Along with the unveiling of the trials machine came the news that the designer, Paul Wright, also planned a scrambler and a road racer.

The press tried the prototype trials model and, although they found some minor snags, in the main they approved of the newcomer. The next step was the first scrambler and this followed the lines of the trials machine to a great extent but was powered by the 34A engine with Marcelle conversion. Most other changes were just to suit the use to which the machine was put, while frame and forks remained the same except for the engine loop. This became duplex with the tubes just far enough apart to allow the exhaust pipe to pass between them.

During 1963 work continued to improve the prototypes and late in the year the company moved into premises in Whitfield in Northamptonshire. For 1964 the frame was modified but remained in spine form with the frame loop bolted in place. A total of five models were listed all with 250 cc engines and all using the same basic chassis. The Trials Type 2 Standard used the 32A engine, while the Replica had a Parkinson conversion. For scrambles the Moto-Cross Mk4 Standard was fitted with the 36A engine, the Replica with the Parkinson conversion and the Special with the Starmaker.

At the end of the year the range for 1965 was announced as two scramblers and a road racer. The latter was completely new with Scorpion designed engine with their own head, barrel, piston, crankcases and chaincase. Dimensions were 66 × 72 mm using an Alpha crankshaft. An Albion five-speed gearbox was specified.

The engine was hung from a tubular spine frame with swinging fork rear end controlled by Girling units and Ceriani-Scorpion telescopics at the front. 200 mm Oldani brakes were fitted to both wheels and the machine was finished off with clip-ons, rear-sets, racing tank, racing seat and rev-counter.

121

The same basic engine was used by the Avenger MX 4 scrambler and this had the same front forks fitted. The Starmaker powered machine became the Scrambles Special still fitted with the leading link forks, while the remainder of the range was no longer listed.

Unfortunately the newly announced range soon vanished as production ceased in the early part of 1965.

Colours

1964: Trials—black. Scrambles with 36A engine—flamboyant blue with black tank and mudguards. Scrambles with Starmaker engine—flamboyant gold with black tank and mudguards.

Sprite

Frank Hipkin and his partner Fred Evans were plumbers by trade before they moved into the retail motorcycle business around 1960. Frank competed successfully in scrambles and being a little dissatisfied with the machine he rode decided to modify it. The result was a Cotton frame with an extra down tube from steering head to engine mounting, to make it a triplex, Norton front forks, Alpha bottom half, Greeves head and barrel, and Villiers four-speed gearbox.

This was the first Sprite and so well did it go that soon Frank was being asked by other riders to build one for them. A bold step for a couple of ex-plumbers but they took it and became motorcycle manufacturers. They decided that they would not skimp their machines but would use quality parts and where they made the items themselves made sure they had skilled workers.

The forks were changed to AMC which were a little lighter and whose split ends suited the Motoloy front hub. The tank and mudguards were in fibreglass as were the side panels for the competition numbers. A large paper air filter was used and the one chosen was a common fit for several popular cars and thus cheap and easy to obtain.

Sprite scrambler in 1965 with under engine exhaust system

The engine remained as in the prototype with the barrel slightly tuned and the compression ratio 12:1. Ignition was by coil and battery with the latter item rubber mounted. Primary drive was by a single strand chain to a three-plate, Ferodo lined clutch all enclosed in an alloy Alpha case.

The machine was finished in silver grey and sold in kit form to save purchase tax. It proved popular and by the middle of 1964 the partners had plans for a new factory, a revised and lighter frame for the scrambler, a prototype trials model and a frame design to take a Triumph 500 unit.

The revised scrambler frame had duplex frame

1967 Sprite trials model with Villiers 32A engine in brief frame

loops with a single, braced top tube, while the engine was changed to run on flywheel magneto ignition. The primary drive was changed to a duplex chain and a shock absorber included in the clutch hub to supplement the one in the rear wheel. The gearbox internal ratios were made closer.

The trials frame was also duplex but much narrower and the engine fitted was the Villiers competition unit with either its standard iron barrel or the Greeves alloy one.

At the beginning of 1965 the front suspension was changed to leading link for both models. The wheel was carried by a loop that passed behind it and its movement was controlled by a pair of Girling units.

By March 1965 the firm was installed in its new factory at Oldbury in Worcestershire and announced that machine kits could be collected from the works at a reduced price. Models were available for trial on a nearby piece of land and both moves were unusual to say the least but most popular with purchasers. Machines could also be bought less front fork for the benefit of those who preferred to use AMC, BSA or Ceriani, all of which would fit without modification.

The trials model had a lighter revised frame in which the top member became two tubes laid side by side from steering head to seat tube.

For 1966 two more models joined the range fitted with the Starmaker engine, one for trials and the other for scrambles. The standard versions with 32A or 36A engines with or without alloy barrel continued. The new machines were given the name Monza and revised frames. The scrambler had a still air chamber supplied via an Austin filter and built into part of the rear mudguard moulding. The trials machine had a very abbreviated seat and rear guard.

The next snag for the small Sprite concern was the drying up of engine supplies. To less determined people this would have spelt the end, but to them the answer was simple—find the best unit going and copy it. Also make it better if you can. Not an easy task but they went ahead to build their version of the 360 Husqvarna and quite a few parts finished up being interchangeable.

The next step was to build for export to the USA under the American Eagle brand name and for that market the engine was enlarged to 84×72 mm and 399 cc. In other respects the machine reflected the late 1960s scrambler and had reverted to telescopic front forks. The gearbox still had four speeds and 6 in. drum brakes were fitted to each wheel. The US model was called the Talon.

Alongside the 400 a trials model with 125 cc Sachs engine was offered and followed in 1970 by

Sprite with ISDT air bottle. Own engine based on Husqvarna unit

a similar size scrambler and a 250 built by reducing the 400 dimensions to 69·5 × 64·5 mm and an actual capacity of 245 cc. Aside from the stroke, the bottom halves were essentially the same as was the chassis. If that was not enough, the firm also sold an enduro-type machine to the States fitted with a 125 cc Zundapp engine unit.

So Frank Hipkin entered the 1970s as managing director of a successful business when so many others were going downhill. His range extended a little to include a trials 400 and, in 1972, a 200 Schoolboy scrambler. Changes came fast and furious in the detail work for as soon as the works machines had proved something to be better it was incorporated into the production line. Thus 1972 saw the introduction of the Fastback model with alloy fuel tank in a choice of $1\frac{1}{2}$, $2\frac{1}{4}$ or 3 gallon capacity, alloy side panels and chrome plated frame.

The next year saw the machines on sale in complete form, rather than as a kit, but in 1974 they were discontinued and Frank's Sprite days were at an end.

Sun

The Sun make was built in the 1920s and then motorcycle production ceased until after the war. It restarted in 1946 with an autocycle of conventional form driven by the Villiers Junior de Luxe engine fitted into a rigid frame with pressed steel forks.

In 1949 this was replaced by a model with the 2F engine and this was joined by a motorcycle fitted with the 1F which also had a rigid frame and girder action forks. The two models continued to the end of 1950 when the autocycle was dropped and two other models joined the 98 cc machine which became the de luxe with tubular link action forks.

The new models were the 122 cc with 10D engine and Challenger de luxe with 197 cc 6E. Both engines were fitted into the same cycle parts and the specification was de luxe with telescopic front forks, plunger rear suspension, rectified lighting, battery and electric horn as standard. A saddle and a rear carrier were fitted and provision could be made for a passenger.

There were a few changes to the frame construction and mudguard width for 1952 but otherwise the three models continued as they were. They were joined in March by a Challenger competition model of 197 cc for trials use. This retained the plunger frame with stronger springs and featured a saddle fitted to a raised mounting frame, Avon trials tyres, alloy mudguards, an Amal carburettor, upswept silencer, raised bars and no lights. Due to material supply problems of the times the prototype could not be followed by production immediately.

1952 Motorcycle model with 1F engine in simple cycle parts

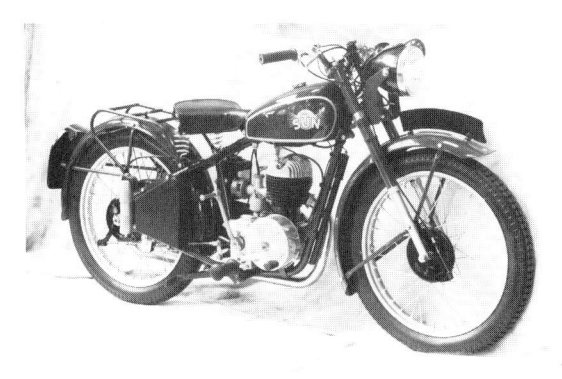

Sun Challenger de luxe in 1952 fully equipped with battery, pump and plungers

For 1953 the smallest model had its engine changed to the 4F unit and the larger ones were given batteries with increased capacity. Otherwise they continued as before and were joined by the competition model. A dualseat was also available for the larger road machines.

During the year there were engine changes to the 122 cc 12D and 197 cc 8E units. For 1954 there were further changes to the use of a frame with swinging fork rear suspension for the 122 and 197 cc models which were joined by the Cyclone model powered by the 1H engine. This used a new frame based on the existing design modified to suit the larger unit. The 122 and 197 models were available with three- or four-speed gearboxes and for the competition model the engine could be supplied in standard or tuned form.

There were further changes in the line up of models for 1955 as the 122 cc machine was replaced by the 30C unit of 147 cc to make the Challenger Mk1A. The 197 cc model became the Challenger Mark IV and the 98 cc the Hornet, and this model was also fitted with a larger fuel tank.

The Cyclone and Competition models continued and the latter was joined by the Scrambler fitted with a tuned engine and three- or four-speed gearbox. The lines were similar to the trials model but front suspension was by Earles forks controlled by Armstrong units.

For 1956 the Hornet continued unaltered as the only model left with a saddle. The 147 Challenger was an export only machine but the 197 was available on the home market and became the Mark V with a larger fuel tank. It was accompanied by a new 197, the Wasp, with 9E engine in a new frame design fitted with side panels which assisted the deeply valanced rear mudguard in protecting the passenger's feet. It was matched at the front by a well valanced guard and the front suspension was by short leading link forks made by Armstrong. A large toolbox, which also housed the battery, was built into the rear panelling under the dualseat.

A competition version of the Wasp was also offered using the same forks but with the usual trials Dunlop saddle, tyres, upswept silencer, alloy guards and raised bars. The Cyclone model continued with detail changes only, while the rigid trials model and the scrambler were dropped.

Late in the year the 1957 range was announced with two new models joining the six existing ones. Of these the Hornet was unchanged apart from its colour, as was the Challenger Mark V which remained in the traditional maroon. The smaller Challenger became the Mark VI with a change of engine to the 31C, while the Cyclone and both Wasps continued with just a silencer improvement on the smaller road machine. Earles or leading link forks were available for the competition machine.

The two new models were the Century and the Wasp Twin. The first used the 8E engine with three-speed gearbox in a Wasp-type frame with a degree of rear enclosure. The second was the Wasp Twin, also called the Overlander, with 2T engine in the Wasp-type frame with enclosure panels and fitted with the leading link forks and a slightly longer wheelbase to accommodate the larger crankcase.

Early in 1957 the Sun Wasp chassis was used as a test-bed for the 2L engine but this never became a production motorcycle although the engine was used by Sun in a scooter. The firm was slightly involved with such machines as it produced one with the 4F engine.

It was a much contracted range for 1958 with the two Challengers, the Cyclone, and the Century all discontinued. The competition Wasp was also dropped but the road model continued with three- or four-speed gearbox available. The Hornet and Overlander also continued but the first was withdrawn early in 1958 with the four-speed Wasp to just leave a two machine range.

This lasted into 1959 when the motorcycles were discontinued to leave the Sun scooters for a year or two before they too went.

Above **Show Sun scrambler in 1955 with 7E engine and Earles forks**

Below **1957–59 Overlander model with 2T engine, leading links and enclosure**

Colours

1951/54: All models—maroon, gold lined tank with Sun transfer, wheel rims silver, chrome plated exhaust system, headlamp rim, controls. Brake pedal and rear plunger boxes silver or dull chrome.

1955: **Cyclone**—black with chrome tank panels. other models: as before.

1956: **Cyclone**—maroon as in 1954.

1957: **Hornet**—Saxe blue with gold lining. **Challenger V**, **VI** and **Cyclone**—maroon. **Wasp** and **Wasp Twin**—Quaker grey with tank panels and tool and battery cases in mushroom. Option of chrome plated tank sides. **Century**—Quaker grey with mushroom tank panels.

1958: **Hornet**—as 1957. **Wasp** and **Wasp Twin** (**Overlander**)—polychromatic silver blue, two-tone fuel tank with silver blue side panels on slightly darker background. Chrome plated tank option.

Sun Challenger 1A with 30C engine in 1955–56

Tandon

The Tandon entered the motorcycle market in 1948 from a factory at Watford in Hertfordshire with a simple lightweight model driven by a Villiers 9D unit. This was the Special, later the Milemaster, Mark I, which carried the twin port engine in a rigid frame with duplex down tubes. Front suspension was by telescopic forks with springs which worked in both compression and tension and the upper ones of which could have their preload adjusted by a handknob at the top of each fork leg.

No rear suspension was fitted but a Lycett saddle was provided and could be adjusted for height. Both brakes were 4 in. diameter and the spoked wheels carried 2·50 × 19 in. tyres.

The engine looked a little old-fashioned for the inlet tract curved round to enter the cylinder on its left side, the plug on the right was matched by the decompressor on the left and the twin exhausts ran from the alloy port adaptors down to a cylindrical silencer carried across the frame ahead of the crankcase. The outlet pipe ran from near the left end of the silencer to pass up and along the top of the primary chaincase to the rear of the machine.

The fuel tank was rather angular in shape and lining with the gear lever attached to its right side and linked to the three-speed gearbox. Lighting was direct and the machine was well equipped with bulb horn. rear carrier and optional speedometer. Unusual was the use of magnesium alloy for a number of parts in the frame and fork assembly and the rear brake pedal. Tools were carried in a channel in the top of the fuel tank and the machine was fitted with a propstand.

In 1950 the Milemaster was joined by another 125 cc model of completely different form. If the first had looked rather pre-war in styling, the second sported a new engine, the 10D, a new frame with swinging fork rear suspension controlled by rubber in compression, and a new finish. The machine looked totally different and a little unusual with the rear fork, wheel, mudguard and stays moving up and down with no apparent support as the rubber, which also damped the movement, was out of sight.

The tank was of a new and more rounded shape but continued to house the tools in a compartment in its top along with the tyre pump and a junction for the electrics. Rectified lighting was fitted with the battery under the saddle so an electric horn was supplied, and the speedometer continued as an option. The front brake remained at its rather inadequate 4 in. diameter and was spongy in operation.

For 1951 the two 125s were joined by the Superglide Supreme with 197 cc 6E engine and 5 in. brakes, the 125 cc Superglide also having the 5 in. rear fitted. There was a detail change to the front wheel spindle fitting to the forks but otherwise the range continued as before.

It was augmented early in 1951 by a competition 125 called the Kangaroo. This continued to use the 10D engine and the rear suspension, albeit stiffened up, at a time when rigid was *de rigueur* for trials. That aside, the machine used suitable tyres, gearing and mudguards.

1952 brought adjustable footrests but otherwise the range was unchanged until late in the year when the 1953 models were listed. The rigid Milemaster was dropped and a 197 cc competition model was introduced as the Kangaroo Supreme. This was given a new swinging fork frame with the fork controlled by a pair of spring units with rubber blocks to act as movement stops. It was fitted with a competition trials seat and competition front plate carrying a 2½ in. diameter Lucas headlight.

The smaller Kangaroo continued fitted with the new style seat, while the smaller Superglide was only changed to use the 5 in. front brake common to the range. The Superglide Supreme had a more conventional form of fuel tank fitted without the top slot for tool storage, this being done with a box that clipped up into the tank tunnel from below.

In place of the Milemaster came a rigid framed

Tandon model built in 1948 for Indian market using 9D engine

1951 Supaglide Supreme with 6E engine and rubber rear suspension

1954 brochure picture of Tandon Twin Supreme with 242 cc Anzani engine

129

125 called the Imp and this was joined by a 197 cc version with conventional swinging fork frame. This used Armstrong units to control the movement and was fitted with a saddle as standard, although a dualseat was available as an option. A neat centre-stand was fitted which had a locking device to prevent it collapsing inadvertently.

At the end of 1953 there were a good number of changes to the range with the 122 cc and 197 cc Superglide and Kangaroo models being dropped. The two Imp models continued but fitted with 12D and 8E engine units, the latter available with a four-speed gearbox in place of the standard three-speed one. The Imp range was extended with a de luxe model which had the four speeds as standard, a larger fuel tank and rectified lighting, and with a Scrambler model. This used a tuned 8E engine, Earles front forks and was equipped to suit its use.

Also new for 1954 were two larger models, one, the Monarch, an Imp Supreme fitted with the 1H engine and the de luxe 2·25 gallon tank, and the other with the same cycle parts but driven by the 242 cc Anzani twin-cylinder engine unit.

There were further changes at the end of the year when the 1955 models were announced. The smallest Imp, the 125, had a further engine change to the 147 cc 30C unit but otherwise remained as before. The two Imp Supreme models were joined by a third called the 'Special' and fitted with Armstrong leading link front forks. In other respects it remained the same and both the Monarch and Twin Supreme were fitted with the leading link forks as standard. In their case the brake size was increased to 6 in. diameter for both wheels and the hubs became full width.

The Scrambler had its engine changed to the sports 7E with four-speed gearbox and was joined by a second version fitted with the 242 cc Anzani engine. This, however, failed to go into production but the twin range was extended by a new model called the Viscount and fitted with the larger 322 cc Anzani unit. In other respects it was a copy of the Twin Supreme with leading link forks and the 6 in. brakes. It had a revised frame also used by the Twin Supreme and the Monarch which followed the lines of the old one.

During 1955 the 197 cc Imp Supreme models were discontinued and at the end of the year the company ceased production for a short while. They restarted in the middle of 1956 with the Imp Supreme Special fitted with the 8E engine and three- or four-speed gearbox in the frame with swinging fork rear and leading link front suspension. Also listed was the Monarch in the same frame but with the 6 in. brakes in place of the 5 in. ones. In this form they continued until all production ceased in the middle of 1959.

It was a good point to stop at for motorcycles were on the point of entering a bleak period and the days of the simple utility machine and its Villiers engine were coming to an end.

Colours
1948: **Milemaster**—black, lined tank, black rims; chrome plated exhaust system, minor controls, lid to toolbox on tank top.
1950: **Milemaster**—as 1948. **Superglide**—metallic blue frame, forks, tank, mudguards, chainguard; black headlamp shell; chrome plated tank top toolbox lid, wheel rims, headlamp rim, exhaust system, bars.
1951/52: **Milemaster**—as 1948, or in polychromatic blue. **Superglide**—as 1950, or in red or black. **Kangaroo**—black frame, forks, chainguard, front part of tank; rear part of tank chrome plated; double white breaker line; alloy mudguards; chrome plated rims.
1953: **Imp**—beige. **Imp Supreme**—beige frame, forks, tank, mudguards, toolboxes, chainguard; red lined wheels; red panel on tank sides.
1954: **Imp** and **Imp Supreme**—as 1953. **Imp Supreme de luxe**—as **Imp Supreme Scrambler**, **Monarch** and **Twin Supreme**—beige.
1955: **Viscount**—beige with red tank flash as **Imp**, option of black with ivory tank panel.

Part Three Appendix

269 cc (70 × 70 mm)		Prefix
Mk I	1913, separate magneto	O
Mk II	1916-20	A
Mk III	1920-21	B
Mk IV	later models with Villiers flywheel magneto	C
Mk V	1922, flywheel magneto	D
147 cc (55 × 62 mm)		
Mk VI-C	1922-23	H
Mk VII-C	1923-24	L
Mk VIII-C	1924, built to end of 1947	W
148 cc (53 × 67 mm)		
Mk XII-C	1932-40	GY, GYF
Mk XV-C	1934-40	CUX, CUXF
172 cc (57·15 × 67 mm)		
1924-32, Sports, petroil		T
1924-32, Sports, autolube		TL
1926-32, Super Sports TT, autolube		BZ
1925-34, Brooklands, autolube		Y
196 cc (61 × 67 mm)		
Mk 1E	1928 to 1938, petroil or autolube	1E
Mk2E	1930-40, petroil only	XZ
Super Sports 1929-40		KZ, KZS
197 cc (59 × 72 mm)		
Mk 3E	1939-40, three speed unit construction	VV
247 cc (67 × 70 mm)		
Mk VI-A	1922-23	J
Mk VII-A	1924-25	S

Postwar

Villiers adopted a system where each engine unit for a particular application and machine manufacture was given a three digit number followed by a serial number. It would appear that this was done for all units whether for motorcycles, lawn mowers, generators or any other purpose so that prefix 999 was reached in about 1951. A letter was then added after the prefix to give 001A and when 999A was reached a B series was started.

Taking the data from the machine numbering produces the following table running from the continued use of wartime letter prefixes up to the final series. There are no C series and the table shows where more than one model or make used the same specification engine and also where Villiers used the same number for more than one engine type.

The data is not complete as often no record could be found to confirm the use of an engine for all the years that it is believed to have been fitted. The years are model years and in general run from the September preceding and are given as a guide.

While incomplete, it is believed that this is the only comprehensive Villiers engine listing outside the factory and it is presented in that light to offer assistance to readers.

Villiers engine numbers

Pre-war system

In this each engine was stamped with one or more prefix letters and a serial number. Not all are dated but years have been included where known. Listed by capacity.

Mk VIII-A	1925-26	X
Mk IX-A	1926-29	DZ
Mk X-A	1930-32	JZ
Mk XVI-A	1934-40	AXF
249 cc (63 × 80 mm)		
Mk XIV-A	1933 to 1940 aircooled, petroil	BYP
	air cooled, autolube	BY
	water cooled, autolube	RY
Mk XVII-A	1935-38	BYX
Mk XVIII-A	1939-40	UU
343 cc (79 × 70 mm)		
Mk VI-B	1922-23	K
MkVII-B	1924-25	M
MkVIIIB	1925-26	AZ
MkIX-B	1926-29	CZ
Mk IX-BA	1929-32	CZA
Mk X-B	1928-32	HZ
346 cc (70 × 90 mm)		
Mk XIV-B	1931-40 autolube	YZ
	1931-40 petroil	YZP
Mk XVII-B	1932-40	WY
98 cc (50 × 50 mm)		
Midget	1931-40	CY, CYA
Junior	1934-39	SH
Junior de luxe	1940-41	XX
	1941-48	XXA
122 cc (50 × 62 mm)		
Mk VIII-D	1936-38 three speeds	AA
Mk 9D	1938-39 three speeds	AAA
Mk 9D	1941-48 extra seal fitted	AAA...A

131

Part Three Appendix

Engine number	Type	Gearbox	Make	Model	Year
XXA	JDL	1	F-B	50	1946–47
			Norman	Autocycle	1946–49
AAA	9D	3	Norman	Motorcycle	1946–48
364	9D	3			
434	JDL	1	F-B	50	1947–49
436	2F	1			
451	JDL	1	Sun	Autocycle	1946–8
563	2T	4	DMW	Dolomite II	1957
597	9D	3	F-B	51	1946–49
632	JDL	1			
662	5E	3	Ambassador	series I & II	1946–48
716	1F	2	BAC	Lilliput	1951
			Bond	Minibyke	1950
			Bown	Motorcycle	1952
			Sun	Motorcycle	1949–51
763	10D	3	OEC	Atlanta S1 & C1	1949–50
797	1F	2	James	J3	1951-53
801	2F	1	Bown	Auto Roadster	1950–52
			F-B	56	1949–51
			Norman	C autocycle	1950–55
			Sun	Autocycle	1949
809	6E	3	Ambassador	series III	1948–49
812	6E	3	OEC	Atlanta S2	1949–50
823	10D	3	F-B	53	1950
824	6E	3	F-B	54	1949
838	10D	3	F-B	52	1949
843	10D	3	Norman	B1	1949–50
844	6E	3	Norman	B2	1949–50
932	10D	3	OEC	D1, S1, C1	1950–51
935	10D	3	F-B	52 & 53	1950–51
938	10D	3	Norman	B1	1951–52
941	10D	3	DMW	125 standard	1951
			Sun	122 de luxe	1951
944	6E	3	Ambassador	Series III & IV	1950–51
945-D	10D	3	DMW	125 de luxe	1951
945	6E	3	OEC	C2, D2, S2	1950–51
946	6E	3	F-B	54 & 55	1950–51
947	6E	3	Norman	B2	1951
949	6E	3	Ambassador	Series III & V	1950–51
983	1F	2	Norman	D	1951–53
036A	10D	3	DMW	125 Comp	1951
139A	6E	3	Ambassador	Supreme	1951–53
158A	6E	3	F-B	60	1952
161A	1F	2	Bond	Minibyke	1951–53
166A	10D	3	F-B	59	1952
171A	2F	1	New Hudson	Autocycle	1952–55
189A	2F	1	F-B	56	1951–52

Engine number	Type	Gearbox	Make	Model	Year
206A	10D	3	F-B	52 & 53	1951–53
207A	6E	3	F-B	54 & 55	1951–53
208A	10D	3	F-B	57	1952–53
209A	8E	3	F-B	58	1952–53
216A	10D	3	OEC	S1, D1, C1	1952–53
218A	10D	3	DMW	125	1952–53
221A	13D	3	Norman	B1S, B1	1953–54
221A	10D	3	Norman	E	1953
227A	6E	3	Ambassador	Popular	1952
228A	6E	3	OEC	S2, D2, ST2	1953
229A	6E	3	Norman	B2, B2S	1952–3
231A	6E	3	Ambassador	Embassy etc	1952–3
257A	10D	3	Tandon	Superglide	1953
262A	10D	3	Sun	10D	1952
271A	6E	3			
319A	7E	4	James	J9	1953
339A	6E	3	Norman	B2C	1953
			Sun	6E	1953
361A	8E	3	Greeves	20R	1954
			OEC	ST2	1954
363A	6E	3	Ambassador	Popular, Embassy, s/c	1953/55
365A	6E	3	James	J7	1953
365A	8E	3	James	K7	1955
374A	7E	4C	F-B	64	1953
375A	7E	4W	F-B	62	1953
380A	7E	4	OEC	ST3	1954
380A	7E	4	James	K7C	1955
381A	7E	4	Norman	B2C	1954–55
381A	7E	4	James	J7C	1953
387A	8E	3	F-B	55 and 67	1953
389A	10D	3	OEC	D55	1954
392A	12D	3	DMW	125	1953
			Sun	12D	1953–54
403A	10D	4C	F-B	63	1953
411A	1H	4	Sun	1H	1954–57
			Tandon	Monarch	1956
419A	6E	3	Sun	6E	1951
420A	10D	4W	F-B	61	1953
489A	6E	3	Sun	6E	1952
491A	8E	4	Greeves	20D, 20S, 20T	1954
521A	4F	2	James	J11	1954-55
534A	4F	2	Norman	D	1954–56
569A	13D	3	James	J5	1953
586A	30C	3	Sun	Challenger	1955–56
A608	JDL	1	Aberdale	Autocycle	1949
618A	13D	3	Norman	B1S	1954
618A	13D	3	James	J5	1955

Villiers Singles & Twins

133

Part Three Appendix

Engine number	Type	Gearbox	Make	Model	Year
652A	4F	2	Excelsior	F4, F4S	1954–56
654A	6E	3	Ambassador	Supreme, s/c	1953–54
659A	7E	4	DMW	Moto Cross Mk 5	1955
669A	12D				1953-54
710A	8E	3	Norman	B2, B2S	1954–55
716A	1F	2	Sun	1F	1952
716A	4F	2	Sun	4F	1953
739A	4F	2	Sun	4F	1954–57
799A	1I1	4	James	K12	1954-55
801A	8E	3	Ambassador	Emb., Sup., Pop.	1954–55
835A	12D				
836A	12D				
842A	1H	4	F-B	68, 71, 75	1954–57
899A	8E	3	Greeves	20R	1955
900A	8E	3	Greeves	20D, 20T	1955
900A	7E	4	Greeves	20S	1955
935A	1H	4	Ambassador	Supreme	1954–57
950A	2T	4	Ambassador	Supreme Twin	1958–59
			Cotton	Villiers Twin	1957–58
			Cotton	Herald	1959
			Greeves	25SA, 25D	1957–60
			Panther	35	1957–59
			Sun	2T	1957–59
993A	1H	4	DMW	Cortina	1955–57
013B	9E	4	Ambassador	Envoy	1956
014	31C	3	Ambassador	Popular	1956
015	31C	4	Ambassador	Popular	1956
018	9E	3	Ambassador	Envoy	1957
019	9E	4	Ambassador	Envoy	1957
021	2T	4	Ambassador	Supreme Twin	1957
030B	30C	3	Ambassador	Popular	1956
040B	8E	3			
043B	7E	4	Norman	B2C	1956
044B	8E	3	DMW	200P Mk 1	1958
045B	8E	3	Ambassador	Envoy, Embassy, s/c	1955
046B	8E	4	Sun	Challenger	1954–56
047B	8E	3	Greeves	20R3	1957
049B	8E	3	Excelsior	R4, R5, R6	1956–57
051B	31C	3	Excelsior	U11	1961
053B	8E	3	DMW	200P	1955–56
055B	8E	4	Sun	Trials	1955
056B	8E	4	Tandon	Imp Supreme	1956
063B	8E	3	Tandon	Imp Supreme	1955
064B	8E	4	Ambassador	Embassy	1956
065B	8E		James	Captain	1957
066B	8E	3	Norman	B2S	1956
074B	8E	3	Greeves	20R3	1956

Engine number	Type	Gearbox	Make	Model	Year
127B	30C	3	Norman	B1S	1955-59
133B	29C	4	DMW	Leda	1955-56
155B	7E	4C	Greeves	20R4	1957-58
164B	8E	4	Greeves	20D	1955
176B	2F	1	New Hudson	Autocycle	1956
178B	2F	1	Norman	C	1956-57
228B	9E	3	Cotton	Vulcan	1958
270B	9E	4	Excelsior	A9	1958
			Greeves	20D	1957
295B	30C	3	F-B	73, 78	1956-59
			James	L15	1958-59
298B	7E	4	James	K7T	1957
310B	4F	2	James	L1	1958
313B	9E	4	Ambassador	Envoy	1958
			Cotton	Vulcan Sports	1961-62
			Excelsior	A9	1956
			Greeves	20D	1958
			Panther	10/4	1956-62
324B	9E	4	Greeves	20D	1956
326B	9E	4	DMW	200P Mk9	1956
331B	9E	4	Greeves	20T	1956
355B	8E	3	Panther	10/3	1960
356B	30C	3	Excelsior	U9, U9R	1959-60
369B	9E	4	Norman	B2SC	1956-57
375B	8E	3	Norman	B2S	1956
377B	9E	4	Sun	Wasp	1956-57
414B / 415B	9E	4	DMW	M-X Mk 6 / Mk 7 Trials	1956-58
437B	9E	4	Sun	Wasp Comp	1956-57
478B	9E	4	Cotton	Trials	1959
504B	33A	4	Cotton	Scrambler	1959
512B	2L	3 and 4	Ambassador	Statesman, Pop.	1958-60
514B	6E	3			
525B	31C	3	Ambassador	Popular	1958-59
538B	31C	4	Norman	B1S	1959
543B	2H	4	Ambassador	Supreme single	1958
544B	9E	3	Cotton	Vulcan	1960
			Panther	10/3A	1957-62
545B	6F	2			
580B	9E	4	Ambassador	3 Star, Envoy	1959-62
602B	4F	2	James	L1	1960
604B	4F	2	Sun	Hornet	1958
605B	4F or 6F	2	Excelsior	F4, F6S, SB1, CA8	1957-58
606B	6F	2	Excelsior	F & C ranges	1959-64
607B	6F	2	James	L1	1959-64
625B	9E	4	Greeves	20TC, TD, TE, DB, SC	1960-62
626B	9E	4	Greeves	20SAS, SCS, DC, SA	1958-60

135

Engine number	Type	Gearbox	Make	Model	Year
634B	2L	4	DMW	175P Mk 9	1957
652B	10E	4	James	K7, K7T	1958–59
662B	10E	3	F-B	81	1958–59
699B	2T	4	Cotton	Continental	1961
723B	9E	4	Greeves	20TA, TAS, TCS	1959–60
	32A	4	Greeves	24TDS	1961
726B	9E	3	Sun	Wasp	1959
732B	9E	3 or 4	Norman	B2S	1959–61
733B	9E	4	Norman	B2SC	1959–60
734B	2T	4	Norman	B3, B4	1958–61
770B	9E	4	Ambassador	Envoy, 3 Star	1959–62
808B	2T	4	Ambassador	Super S	1959–61
810B	2T (E)	4	Panther	35	1966
847B	3T	4	Cotton	Messenger	1959
			Panther	45, 50	1959–60
862B	2T	4			
863B	9E	4	Cotton	Trials	1960
863B	32A	4	Greeves	24TCS, TD, TE	1960–62
864B	32A	4	Greeves	24DB, DC	1959–62
888B	31A	4	Greeves	24SAS	1959
893B	3T	4	Greeves	32DB	1960
			Panther	45, 50	1960–61
004D	2T	4	Cotton	Herald, Double Glos.	1962–63
			Panther	35, 35S	1961–62
016D	33A	4	Cotton	Scrambler	1960
039D	2T	4	Cotton	Herald, Double Glos.	1960–61
040D	9E	4	Cotton	Vulcan	1965
051D	31C	3	Excelsior	U10, U14	1960–65
053D	3T	4	Panther	50	1961–62
054D	3T	4	Cotton	Messenger	1960–62
059D	9E	4	Excelsior	R10, R11, ER11	1960
070D	33A	4	Cotton	Scrambler	1961
			Greeves	24SCS, MCS	1960–61
149D	32A	4	Cotton	Trials	1961–63
150D	9E	4	Cotton	Trials	1961–62
161D	32A	4	DMW	Mk 15 Trials	1962
208D	2T	4	Ambassador	Electra 75	1961
			Panther	35	1967–68
214D	2T	4	Ambassador	Super S	1962
221D	2T	4	Greeves	25DC	1961
222D	3T	4	Greeves	32DC, DCX, DD	1961–64
229D	2T	4	Cotton	Continental Sports	1963
248D	34A	4	Cotton	Scrambler	1962
251D	9E	4	Cotton	Vulcan	1966
			Greeves	20DC, TE	1962–66
269D	2T	4	Ambassador	Super Sports	1961–62

Engine number	Type	Gearbox	Make	Model	Year
284D	34A	4	Cotton	Cougar	1962–63
			DMW	Mk 14 scrambler	1964
			Greeves	24SC, MDS	1962
305D	34A	4	Greeves	24MCS	1962
393D	2T	4	Cotton	Continental	1962
400D	31A	4	Cotton	Corsair	1962
428D	2T	4	James	M25	1962–63
429D	2T	4	F-B	89	1962
453D	32A	4	Greeves	24TD, TES, TFS	1963–65
500D	9E	4	F-B	89	1962
507D	9E	3	Ambassador	Popular	1962
606D	2T	4	Cotton	Continental	1963
607D	32A	4	Cotton	Trials	1963–66
718D	2T	4	Greeves	25DC, DCX, DD	1962–64
725D	2T	4	Cotton	Herald	1963
730D	6F	2	Excelsior	F12, C14	1962–63
757D	Starmaker	4	Cotton	Cobra	1963
			Greeves	24ME	1963–64
004E	9E	3 or 4	Cotton	Vulcan	1963
085E	36A	4	Greeves	24RAS, MDS	1964–65
086E	36A	4	Cotton	Cougar	1963–66
			Greeves	24MD, MDS	1964
146E	9E	3	Cotton	Vulcan	1957
296E	36A	4	Greeves	24RAS	1963–64
326E	32A	4W	F-B	92	1964–65
			James	M25T	1964–65
352E	36A	4C	F-B	93	1964
			James	M25R	1964–65
353E	3T	4	Cotton	Messenger	1964
353E	4T	4	Cotton	Continental	1965–66
			Greeves	25DC, DCX, DD	1964–65
490E	Starmaker	4	Cotton	Telstar	1964
687E	4T	4	Cotton	Continental	1968
			F-B	89, 91	1964–65
			James	M25S	1964–65
688E	4T	4	F-B	89	1965
826E	Starmaker	4	Cotton	Cobra	1965–68
			F-B	94	1964–65
			James	M25RS	1964–65
834E	Starmaker	4	Cotton	Telstar	1964–66
863E	31A	4	Cotton	Corsair	1964
871E	Starmaker	4	Cotton	Trials	1965
972E	Starmaker	4	Cotton	Conquest	1965–68
086F	4T	4	Greeves	25DC	1966
131F	Starmaker	4	Cotton	Trials	1966–68
161F	37A	4	Cotton	Trials 37A	1967–68
			Greeves	24TGS, THS, TJS, TJ	1966–69

Villiers Singles & Twins

137

Starmaker engines

These units were built to various specifications to suit different applications as set out below:

Specification	Description
757D	Standard scrambles specification, with twin Amal Monobloc carburettors
490E	Roadracing version of Specification 757D, but with close-ratio gears and tachometer drive
494E	Scrambler, as Specification 757D, but with 6-volt direct lighting
826E	Scrambler, with single Amal Monobloc carburettor
827E	Scrambler, with single Amal Monobloc carburettor and 6-volt direct lighting
834E	Road racer, with close-finned cylinder and head, close-ratio gears, tachometer drive, outrigger contact breaker and Amal GP2 carburettor
865E	Road racer, as Specification 834E, with 6-volt direct lighting
871E	Trials engine, with wide-ratio gears and Villiers Type S.25 carburettor, but without lighting
972E	Sports roadster, with single Amal Monobloc carburettor and 12-volt rectified lighting
091F	Special, with single Amal Monobloc carburettor, but without clutch, gearbox and primary drive
131F	Trials engine with 6-volt direct lighting

Engine specifications

Specification for Villiers

Model	JDL	5E		
Year from	**1946**	**1946**		
Year to	**1948**	**1948**		
Bore (mm)	50	59		
Stroke (mm)	50	72		
Capacity (cc)	98	197		
Carb. make	Villiers			
Carb type	Junior			
Box ratio: top		1·0		
Box ratio: 2nd		1·6		
Box ratio: 1st		2·86		

Model	IF	2F	4F	6F
Year from	**1948**	**1948**	**1952**	**1956**
Year to	**1953**	**1958**	c. **1958**	c. **1964**
Bore (mm)	47	47	47	47
Stroke (mm)	57	57	57	57
Capacity (cc)	99	99	99	99
Carb. make	Villiers	Villiers	Villiers	Villiers
Carb. type	6/0	Junior **1**	6/0 **1**	6/0 **1**
Comp. ratio (:1)	8·0	8·0	8·0	8·0

Power: bhp	2·8	2·0	2·8	2·8
@ rpm	4000	3750	4000	4000
Inlet angle °	126	126	126	
Exhaust angle °	118	118	118	
Transfer angle °	100	100		
Ignition time (in.)	0·125	0·125	0·125	0·125
Points gap (in.)	0·015	0·015	0·015	0·015
Front chain pitch	$\frac{3}{8} \times \frac{5}{32}$	$\frac{3}{8} \times \frac{5}{32}$	$\frac{3}{8} \times \frac{5}{32}$	$\frac{3}{8} \times \frac{5}{32}$
Rear chain pitch	$\frac{1}{2} \times \frac{3}{16}$	$\frac{1}{2} \times \frac{3}{16}$	$\frac{1}{2} \times \frac{3}{16}$	$\frac{1}{2} \times \frac{3}{16}$
Sprocket: engine (T)	17	17	17	17
Sprocket: clutch (T)	42	42	42	42
Sprocket: gearbox (T)	14	11	14	14

Lots of James Comets awaiting delivery in 1956

Box ratio: top	1·0		1·0	1·0
Box ratio: 1st	1·54		1·54 **2**	1·72

1 1957—S12 **2** 1953—1·64, 1956—1·72

Model	**9D**	**10D**	**11D**	**12D/13D**
Year from	**1938**	**1948**	**1953**	**1953**
Year to	**1948**	**1953**	**1954**	**1954**
Bore (mm)	50	50	50	50
Stroke (mm)	62	62	62	62
Capacity (cc)	122	122	122	122
Carb. make	Villiers	Villiers	Villiers	Villiers
Carb. type	L/W **1**	3/4	S24	S19
Comp. ratio (:1)	6·5	8·25	10·0	8·25
Power: bhp	3·15	4·8	6·1	4·9
@ rpm	4000	4400	4600	4400
Inlet angle	117	120		120
Exhaust angle	130	140		140
Transfer angle degree	117	108		108
Ignition time (in.)	0·312	0·156	0·156	0·156
Points gap (in.)	0·015	0·015	0·015	0·015
Front chain pitch	3/8	3/8	3/8	3/8
Rear chain pitch	1/2	1/2	1/2	1/2
Sprocket: engine (T)	19	18	18	18
Sprocket: clutch (T)	44	51	51	51
Sprocket: gearbox (T)	12	15	15 or 17	15 or 17
Box ratio: top			1·0	
Box ratio: 3rd			1·35	
Box ratio: 2nd			2·3 1·8	
Box ratio: 1st			3·47 2·93	
Box ratio: top	1·0	1·0	1·0	1·0
Box ratio: 2nd	1·625	1·7 **2**	1.34 1·7	1·34
Box ratio: 1st	2·925	3·25 **2**	2·55 3·25	2·55

1 WD—3/1 **2** 1950 on—1·4, 2·66

Model	**29C/4**	**30C**	**31C**	**2L**
Year from	**1954**	**1954**	**1956**	**1956**
Year to	**1956**	**1959**	**1964**	**1962**
Bore (mm)	55	55	57	59
Stroke (mm)	62	62	58	63·5
Capacity (cc)	147	147	148	174
Carb. make	Villiers	Villiers	Villiers	Villiers
Carb. type	S25	S19	S19	S22
Comp. ratio (:1)	10·0	8·25	7·75	7·4
Power: bhp	6	5·4	6·3	7·4
@ rpm	4500	4300	5000	5000

Inlet angle degree	128	128	127	141
Exhaust angle degree	156	156	151	149
Transfer angle degree	120	120	126	143
Ignition time (in.)	0·156	0·156	0·172	0·172
Points gap (in.)	0·015	0·015	0·015	0·015
Front chain pitch	$\frac{3}{8}$	$\frac{3}{8}$	$\frac{3}{8}$	$\frac{3}{8}$
Rear chain pitch	$\frac{1}{2}$	$\frac{1}{2}$	$\frac{1}{2}$	$\frac{1}{2}$
Sprocket: engine (T)	23	23	20	20
Sprocket: clutch (T)	51	51	43	43
Sprocket: gearbox (T)	15	15	16	17
Box ratio: top	1·0		1·0	1·0
Box ratio: 3rd	1·35		1·27 1·35	1·27
Box ratio: 2nd	2·3		1·78 2·4	1·78
Box ratio: 1st	3·47		2·94 3·6	2·94
Box ratio: top	1·0	1·0	1·0	1·0
Box ratio: 2nd	1·34	1·34	1·34 1·7	1·34
Box ratio: 1st	2·55	2·55	2·55 3·25	2·55

Model	**6E**	**7E**	**8E**	**9E**
Year from	**1948**	**1953**	**1953**	**1955**
Year to	**1953**	**1956**	**1958**	**c. 1967**
Bore (mm)	59	59	59	59
Stroke (mm)	72	72	72	72
Capacity (cc)	197	197	197	197
Carb. make	Villiers	Villiers	Villiers	Villiers
Carb. type	4/5	S24	S24 or S25	S25/1
Comp. ratio (:1)	8·0	8·25	7·25	7·25 **1**
Power: bhp	7·5 **2**	9·3	8·4	8·4 **3**
@ rpm	4000	4300	4000	4000 **3**
Inlet angle °	134		134	134
Exhaust angle °	152		152	152
Transfer angle degree	126		126	126
Ignition time (in.)	0·156	0·156	0·156	0·172
Points gap (in.)	0·015	0·015	0·015	0·015
Front chain pitch	$\frac{1}{2}$	$\frac{3}{8}$	$\frac{3}{8}$	$\frac{3}{8}$
Rear chain pitch	$\frac{1}{2}$	$\frac{1}{2}$	$\frac{1}{2}$	$\frac{1}{2}$
Sprocket: engine (T)	19	23	23	20
Sprocket: clutch (T)	38	51	51	43
Sprocket: gearbox (T)	15	17 **4**	17 or 19 **5**	18
Box ratio: top		1·0	1·0	1·0
Box ratio: 3rd		1·35	1·35	1·27 1·34
Box ratio: 2nd		1·8 2·3	1·8 2·3	1·78 2·4
Box ratio: 1st		2·93 3·47	2·93 3·47	2·94 3·6
Box ratio: top	1	1·0	1·0	1·0
Box ratio: 2nd	1·7 **6**	1·34 1·7	1·34 1·7	1·34 1·7

Villiers Singles & Twins

141

Part Three Appendix

| Box ratio: 1st | 3·25 **6** | 2·55 3·25 | 2·55 3·25 | 2·55 3·25 |

1 comp—8·25 **2** 1951—8·2 **3** comp—9·3/4300 **4** 15 option for 4 speed **5** 17 std with 4 speed, option—15 **6** 1950 on—1·4, 2·66

Model	**1H**	**2H**	**31A**	**32A**
Year from	**1953**	**1957**	**1958**	**1960**
Year to	**1957**	**1960**	**1960**	**1965**
Bore (mm)	63	66	66	66
Stroke (mm)	72	72	72	72
Capacity (cc)	224	246	246	246
Carb. make	Villiers	Villiers	Villiers	Villiers
Carb. type	S25	S25/5	S25/1	S25/5
Comp. ratio (:1)	7·0	7·25	7·4	7·9
Power: bhp	10	11		12·4
@ rpm	4500	4500		5000
Inlet angle degree	136	142		
Exhaust angle degree	144	150		
Transfer angle degree	118	118		
Ignition time (in.)	0·172 **1**	0·172	0·172	0·172
Points gap (in.)	0·015	0·015	0·015	0·015
Front chain pitch	⅜	⅜	⅜	⅜
Rear chain pitch	½	½	½	½
Sprocket: engine (T)	20	20	20	20
Sprocket: clutch (T)	43	43	43	43
Sprocket: gearbox (T)	18 **2**	18	15 to 20	15 to 20
Box ratio: top	1·0	1·0	1·0	1·0
Box ratio: 3rd	1·32	1·32	1·56	1·27
Box ratio: 2nd	1·9	1·9	2·18	1·78
Box ratio: 1st	3·06	3·06	3·6	2·94

1 1955—0·156 **2** option 17 or 19

Model	**33A**	**34A**	**36A**	**37A**
Year from	**1958**	**1960**	**1962**	**1965**
Year to	**1959**	**1962**		**1968**
Bore (mm)	66	66	66	66
Stroke (mm)	72	72	72	72
Capacity (cc)	246	246	246	246
Carb. make	Amal	Amal	Amal	Villiers
Carb. type	389	389	389	S25/5
Comp. ratio (:1)	12·0	12·0	12·0	7·9
Power: bhp		19		12·5
@ rpm		6000		5000
Inlet angle degree				140
Exhaust angle degree				150
Transfer angle degree				132

Ignition time (in.)	0·140	0·140	0·140	0·170
Points gap (in.)	0·015	0·015	0·015	0·015
Front chain pitch	3/8	3/8	3/8	3/8
Rear chain pitch	1/2	1/2	1/2	1/2
Sprocket: engine (T)	20	20	20	20
Sprocket: clutch (T)	43	43	43	43
Sprocket: gearbox (T)	15 to 20	15 to 20	15 to 20	15 to 20
Box ratio: top	1·0	1·0	1·0	1·0
Box ratio: 3rd	1·27	1·27	1·27	
Box ratio: 2nd	1·78	1·78	1·78	
Box ratio: 1st	2·55	2·55	2·55	

Model	**Starmaker**	**2T**	**3T**	**4T**
Year from	**1962**	**1956**	**1958**	**1963**
Year to	**c. 1969**	**1968**	**1964**	**1967**
Bore (mm)	68	50	57	50
Stroke (mm)	68	63·5	63·5	63·5
Capacity (cc)	247	249	324	249
Carb. make	Amal	Villiers	Villiers	Villiers
Carb. type	various	S22/2	S25/3	S25/6
Comp. ratio (:1)	12·0	8·2	7·25	8·75
Power: bhp	25 **1**	15	16·5	17
@ rpm	6500 **1**	5500	5000	6000
Inlet angle degree	136 **2**	131	131	
Exhaust angle degree	172 **2**	150	150	
Transfer angle degree	126 **2**	111	111	
Ignition time (in.)	0·095	0·187	0·187	0·155
Points gap (in.)	0·015	0·015	0·015	0·015
Front chain pitch	3/8 duplex	3/8	3/8	3/8
Rear chain pitch	1/2	1/2	1/2	1/2
Sprocket: engine (T)	20	20	25 **3**	20
Sprocket: clutch (T)	43	43	43	43
Sprocket: gearbox (T)	15 or 18	18	16	
Box ratio: top	1·0 **4**	1·0	1·0	1·0
Box ratio: 3rd	1·25	1·32 1·4	1·32 1·4	1·32
Box ratio: 2nd	1·66	1·9 2·02	1·9 2·02	1·9
Box ratio: 1st	2·52	3·06 3·6	3·06 3·6	3·06

1 racing—32/8000   **2** racing—152, 176, 132   **3** to engine 035D/523—20   **4** also wide and racing close

Specification for AJS

Model	Y4 & FB250	Y5 & FB360	Y51
Year from	**1970**	**1970**	**1971**
Year to	**1982**	**1982**	**1974**
Bore (mm)	68	83	83
Stroke (mm)	68	68	74
Capacity (cc)	247	368	400
Carb. make	Amal	Amal	Amal
Carb. type	932	1034	1034
Comp. ratio (:1)	11·0	9·85	10·75
Inlet angle degree	152	170	154
Exhaust angle degree	180	180	180
Transfer angle degree	135	130	130
Ignition time (in.)	0·098	0·098	0·126
Points gap (in.)	0·015	0·015	0·015
Front chain pitch	$\frac{3}{8}$ duplex	$\frac{3}{8}$ duplex	$\frac{3}{8}$ duplex
Rear chain pitch	$\frac{5}{8}$	$\frac{5}{8}$	$\frac{5}{8}$
Sprocket: engine (T)	20	24	24
Sprocket: clutch (T)	43	40	40
Sprocket: gearbox (T)	13	13	13
Box ratio: top	1·0	1·0	1·0
Box ratio: 3rd	1·25	1·25	1·25
Box ratio: 2nd	1·51 or 1·66	1·51 or 1·66	1·51 or 1·66
Box ratio: 1st	2·00 or 2·52	2·00 or 2·52	2·00 or 2·52

Specification for AMC

Model	150	175	200	250
Year from	**1959**	**1957**	**1959**	**1956**
Year to	**1966**	**1960**	**1966**	**1963**
Bore (mm)	55	59	59	66
Stroke (mm)	62·7	62·7	72·8	72·8
Capacity (cc)	149	171	199	249
Carb. make	Amal	Amal		Amal
Carb. type	375	370		389
Comp. ratio (:1)	7·0	8·25	8·5	8·25 **1**
Ignition time (in.)	0·093			0·25
Points gap (in.)	0·018			
Front chain pitch	$\frac{3}{8}$	$\frac{3}{8}$		$\frac{3}{8}$
Rear chain pitch	$\frac{1}{2}$	$\frac{1}{2}$		$\frac{1}{2}$
Sprocket: engine (T)	18			20
Sprocket: clutch (T)	43			43
Sprocket: gearbox (T)				19
Box ratio: top	1·0	1·0	1·0	1·0
Box ratio: 3rd	1·27	1·3	1·32	1·3 or 1·44
Box ratio: 2nd	1·78	1·85	1·85	1·85 or 2·42

Box ratio: 1st	2·94	2·92	2·94	2·95 or 3·57
Box ratio: top	1·0			
Box ratio: 2nd	1·33 **2**			
Box ratio: 1st	2·57			

1 M-X—10·5, trials—9·25   **2** later—1·46

Specification for Anzani

Model	**250**	**325**			
Year from	**1953**	**1954**			
Year to	**1960**	**1960**			
Bore (mm)	52	60	Transfer angle degree	124	124
Stroke (mm)	57	57	Ignition time degree	28	28
Capacity (cc)	242	322	Front chain pitch	$\frac{3}{8}$ duplex	$\frac{3}{8}$ duplex
Carb. make		Amal	Rear chain pitch	$\frac{1}{2}$	$\frac{1}{2}$
Carb. type		376/38	Sprocket: gearbox	15	
Comp. ratio (:1)	6·5 **1**	6·5 **1**	Box ratio: top	1·0	
Power: bhp	10	15	Box ratio: 3rd	1·35	
@ rpm		4800	Box ratio: 2nd	1·82	
Inlet angle degree	120	120	Box ratio: 1st	2·93	
Exhaust angle degree	144	144			

1 1956—7·5

Specification for Excelsior engines

Model	**Spryt**	**Goblin**	**Spryt II**	**Goblin 125**
Year from	**1946**	**1947**	**1947**	**1948**
Year to	**1948**	**1956**	**1956**	**1949**
Bore (mm)	50	50	50	56
Stroke (mm)	50	50	50	50
Capacity (cc)	98	98	98	123
Carb. make	Amal	Amal	Amal	Amal
Carb. type	259	259	259	
Comp. ratio (:1)		6·6	6·6	
Power: bhp		2·6	2·6	
@ rpm		3500	3500	
Ignition time (in.)		0·187		
Points gap (in.)	0·012	0·012 **1**		
Front chain pitch	$\frac{3}{8}$	$\frac{3}{8}$	$\frac{3}{8}$	$\frac{3}{8}$
Rear chain pitch	$\frac{1}{2}$	$\frac{1}{2}$	$\frac{1}{2}$	$\frac{1}{2}$
Sprocket: engine (T)		17	17	17
Sprocket: clutch (T)		44	44	44
Box ratio: top		1·0		1·0
Box ratio: 1st		1·74		1·74

1 Wipac—0·02

Model	**150**	**Twin—243**	**Twin—328**
Year from	**1952**	**1949**	**1957**
Year to	**1962**	**1962**	**1962**
Bore (mm)	55	50	58
Stroke (mm)	62	62	62
Capacity (cc)	147	243	328
Carb. make	Amal	Amal **1**	Amal **2**
Comp. ratio (:1)	7·8	7·7	7·8
Power: bhp	6·5	10 **3**	18
@ rpm	4000	4500 **3**	5000
Ignition time (in.)	0·156	0·156	
Points gap (in.)		0·015	0·015
Front chain pitch	³⁄₈	½ **4**	³⁄₈
Rear chain pitch	½	½	½
Sprocket: engine (T)		26	26
Sprocket: clutch (T)		54	54
Sprocket: gearbox (T)		17	17
Box ratio: top		1·0	1·0
Box ratio: 3rd		1·35	1·35
Box ratio: 2nd		1·8	1·8
Box ratio: 1st		2·92	2·92
Box ratio: top	1·0		
Box ratio: 2nd	1·48		
Box ratio: 1st	2·64		

1 Sports—2 carbs   **2** twin   **3** 1954—12·1/5000   **4** 1952—³⁄₈ duplex

Specification for JAP

Model	**125**
Year from	**1947**
Year to	**1952**
Bore (mm)	54·2
Stroke (mm)	54
Capacity (cc)	125
Carb. make	Amal
Carb. type	223
Power: bhp	3·5
@ rpm	4000
Points gap (in.)	0·015
Front chain pitch	³⁄₈
Rear chain pitch	½
Sprocket: engine (T)	17
Sprocket: clutch (T)	50
Sprocket: gearbox (T)	14
Box ratio: top	1·0
Box ratio: 2nd	1·6
Box ratio: 1st	2·9

Machine specifications

Aberdale

Model	**Autocycle**
Year from	**1947**
Year to	**1949**
Engine make	Villiers
Engine type	JDL
Gearbox type	1
Front brake dia. (in.)	4·0
Rear brake dia. (in.)	4·0
Front suspension	girder
Rear suspension	rigid
Petrol tank (Imp. gal)	1·5

ABJ

Model	**Autocycle**	**Motorcycle**
Year from	**1949**	**1949**
Year to	**1952**	**1952**
Engine make	Villiers	Villiers
Engine type	2F	1F
Gearbox type	1	2
O/A ratio: top		8·47
Front tyre (in.)	2·25 × 26	2·25 × 26
Rear tyre (in.)	2·25 × 26	2·25 × 26
Front brake dia. (in.)	3·75	3·75
Rear brake dia. (in.)	4·0	4·0
Front suspension	teles	teles
Rear suspension	rigid	rigid
Petrol tank (Imp. gal)	1·5	1·5
Wheelbase (in.)	49·75	49·75
Dry weight (lb)	140	140

AJS

Model	**37 A-T**	**Y4**	**Y4**	**Y5**	**Y51**	**FB250**
Year from	**1969**	**1969**	**1970**	**1970**	**1971**	**1975**
Year to	**1969**	**1969**	**1974**	**1974**	**1974**	**1982**
Engine make	Villiers	Villiers/AJS	AJS	AJS	AJS	AJS
Engine type	37A	Starmaker based	Stormer 250	Stormer 370	Stormer 410	250

Gearbox type	4	4	4	4	4	4
O/A ratio: top	8·1	optional	optional	optional	optional	optional
Front tyre (in.)	2·75 × 21	2·75 × 21	2·75 × 21	2·75 × 21	2·75 × 21	
Rear tyre (in.)	4·00 × 18	4·00 × 18	4·00 × 18	4·00 × 18	4·00 × 18	
Front brake dia. (in.)	6	5	5	5	5	
Rear brake dia. (in.)	6	5	5	5	5	
Front suspension	teles	teles	teles	teles	teles	teles
Rear suspension	s/a	s/a	s/a	s/a	s/a	s/a
Petrol tank (Imp. gal)	1·75	2·0	2·0	2·0	2	2
Wheelbase (in.)	51·5	55·5	54·5	54·5	55·5	55·5
Ground clear. (in.)	9·5	8·5	9	9	9·5	10
Seat height (in.)	30	30	31	31	30	36
Dry weight (lb)	212	220	218	221	229	220

AJS

Model	FB250-MX	FB360-MX	RD360E
Year from	**1975**	**1975**	**1981**
Year to	**1982**	**1982**	**1982**
Engine make	AJS	AJS	AJS
Engine type	250	370	370
Gearbox type	4	4	4
O/A ratio: top	optional	optional	optional
Front suspension	teles	teles	teles
Rear suspension	s/a	s/a	s/a
Petrol tank (Imp. gal)	1·5	1·5	
Wheelbase (in.)	58·5	58·5	
Ground clear. (in.)	12	12	
Seat height (in.)	38	38	
Dry weight (lb)	230	230	215

AJW

Model	125
Year from	**1952**
Engine make	JAP
Engine type	125
Gearbox type	3
Front suspension	teles
Rear suspension	s/a

Ambassador

Model	Series I 1	Series III 2	Popular	Series V 3	Supreme	Sidecar
Year from	**1946**	**1948**	**1950**	**1950**	**1951**	**1952**
Year to	**1948**	**1951**	**1955**	**1955**	**1953**	**1955**
Engine make	Villiers	Villiers	Villiers	Villiers	Villiers	Villiers
Engine type	5E	6E	6E **4**	6E **5**	6E	6E **5**
Gearbox type	3	3	3	3 **6**	3	3
O/A ratio: top	6·22		5·86 **7**	5·86 **7**	5·86	7·20 **8**
Front tyre (in.)	3·00 × 19	3·00 × 19	3·00 × 19	3·00 × 19	3·00 × 19 **9**	3·00 × 19
Rear tyre (in.)	3·00 × 19	3·00 × 19	3·00 × 19	3·00 × 19	3·00 × 19 **9**	3·00 × 19
Front brake dia. (in.)	5	5	5	5	5 **10**	5
Rear brake dia. (in.)	5	5	5	5	5 **10**	5

Front suspension	girder	girder	girder **11**	teles	teles	girder
Rear suspension	rigid	rigid	rigid	rigid **12**	plunger	rigid
Petrol tank (Imp. gal)	2·5	2·1	2·1	2·1	2·1	2·1
Wheelbase (in.)			47	47	47	46
Ground clear. (in.)			6	5	5·75	5
Seat height (in.)	28		29·5	29·5	28·5	29·5
Dry weight (lb.)	185		196 **13**	199 **14**	213	324

1 1948—Series II   **2** 1951—Courier   **3** 1951—Embassy   **4** 1954—8E   **5** 1954—8E   **6** 1955—or 4   **7** 1955—5·74   **8** 1955—7·0   **9** 1953—3·25 × 18   **10** 1953—6   **11** 1955—teles   **12** 1953—plunger   **13** 1955—181   **14** 1955—215

Model	Supreme	Envoy	Popular	Envoy	Supreme	Statesman **1**
Year from	1953	1954	1955	1955	1956	1957
Year to	1958	1958	1958	1959	1958	1960
Engine make	Villiers	Villiers	Villiers	Villiers	Villiers	Villiers
Engine type	1H **2**	8E	30C **3**	9E	2T	2L
Gearbox type	4	3	3 **4**	4 **5**	4	3 or 4
O/A ratio: top	6·09	5·74	5·7	5·74	6·2	6·6
Front tyre (in.)	3·25 × 18	3·00 × 19 **6**	3·00 × 18	3·25 × 17 **7**	3·25 × 17	3·00 × 18
Rear tyre (in.)	3·25 × 18	3·00 × 19 **6**	3·00 × 18	3·25 × 17 **7**	3·25 × 17	3·00 × 18
Front brake dia. (in.)	6	5 **8**	5	6	6	5
Rear brake dia. (in.)	6	5 **8**	5	6	6	5
Front suspension	teles	teles	teles	teles	teles	teles
Rear suspension	s/a	s/a	s/a	s/a	s/a	s/a
Petrol tank (Imp. gal)	2·6	2·6	2·1	2·6	2·75 **9**	2·5 **10**
Wheelbase (in.)	48·75	48·75	47	48·75	49	49
Ground clear. (in.)	7	7	7	7	7·5	5
Seat height (in.)	30	30	30	30	32	28
Dry weight (lb)	238	230	181	215	283	216

1 1958—Popular   **2** 1956—2H   **3** 1956—31C   **4** 1957—or 4   **5** 1958—or 3   **6** 1958—3·25 × 17   **7** from 1958   **8** 1955—6   **9** 1957—3·25   **10** 1958—3·5

Model	Super S	3 Star Special	Electra 75	Sports Super S **1**	Popular
Year from	1958	1959	1960	1961	1962
Year to	1964	1964	1964	1962	1962
Engine make	Villiers	Villiers	Villiers	Villiers	Villiers
Engine type	2T **2**	9E	2T **2**	2T	9E
Gearbox type	4	3 or 4 **3**	4	4	3
O/A ratio: top	5·8 **4**	6·2	5·8 **4**	5·8	6·2
Front tyre (in.)	3·25 × 17 **5**	3·25 × 17 **5**	3·25 × 17 **5**	3·25 × 17	3·00 × 18
Rear tyre (in.)	3·25 × 17 **5**	3·25 × 17 **5**	3·25 × 17 **5**	3·25 × 17	3·00 × 18
Front brake dia. (in.)	7 **6**	6	7 **6**	7	5
Rear brake dia. (in.)	7 **6**	6	7 **6**	7	5
Front suspension	teles	teles	teles	teles	teles
Rear suspension	s/a	s/a	s/a	s/a	s/a
Petrol tank (Imp. gal)	3·5	3·5	3·5	3·5	3·5
Wheelbase (in.)		51			
Ground clear. (in.)		6·5			
Seat height (in.)		31			
Dry weight (lb)	312	262	318	298	249

1 1962—Sports Twin   **2** 1964—4T   **3** 1963—4 only   **4** 1964—6·2   **5** 1963—3·25 × 18   **6** 1963—6

Villiers Singles & Twins

Part Three Appendix

BAC

Model	**Lilliput**	**Lilliput**
Year from	**1951**	**1951**
Year to	**1952**	**1951**
Engine make	Villiers	JAP
Engine type	1F	125
Gearbox type	2	3
O/A ratio: top	4·94	
Front tyre (in.)	2·00 × 20	2·00 × 20
Rear tyre (in.)	2·00 × 20	2·00 × 20
Front brake dia. (in.)	3·5	3·5
Rear brake dia. (in.)	3·5	3·5
Front suspension	teles	teles
Rear suspension	rigid	rigid
Petrol tank (Imp. gal)	1·5	1·5
Wheelbase (in.)	45	45
Ground clear. (in.)	5	5
Seat height (in.)	25·5	25·5
Dry weight (lb)	89	

Bond

Model	**Minibyke**	**Minibyke de luxe**
Year from	**1950**	**1951**
Year to	**1953**	**1953**
Engine make	Villiers	JAP
Engine type	1F	125
Geabox type	2	3
O/A ratio: top	4·94	4·8
Front tyre (in.)	4·00 × 16	4·00 × 16
Rear tyre (in.)	4·00 × 16	4·00 × 16
Front brake dia. (in.)	4	4
Rear brake dia. (in.)	4	4
Front suspension	rigid **1**	teles
Rear suspension	rigid	rigid
Petrol tank (Imp. gal)	1·5	1·5
Wheelbase (in.)	46	46
Ground clear. (in.)	5	5
Seat height (in.)	26·5	26·5
Dry weight (lb)	91 **2**	94 **3**

1 teles from July 1950 **2** 1951—96 **3** 1951—100

Bown

Model	**Auto Roadster**	**98**	**TT**
Year from	**1950**	**1951**	**1952**
Year to	**1954**	**1954**	**1954**
Engine make	Villiers	Villiers	Villiers
Engine type	2F	1F **1**	10D
Gearbox type	1	2	3
O/A ratio: top		8·47	7·55
Front tyre (in.)	2·25 × 21	2·50 × 19	3·00 × 19
Rear tyre (in.)	2·25 × 21	2·50 × 19	3·00 × 19
Front brake dia. (in.)	4·0	4·0	5·0

Rear brake dia. (in.)	4·0	4·0	5·0	
Front suspension	girder	teles	teles	
Rear suspension	rigid	rigid	rigid	
Petrol tank (Imp. gal)	1·5	1·25	2·25	
Wheelbase (in.)	48·5	46·9	46	
Ground clear. (in.)		4·75	6	
Seat height (in.)		28	29·5	
Dry weight (lb)	120	135	178	

1 1953—4F

Butler

Model	**Trials**	**Scrambler**	**Tempest**	**Fury**
Year from	**1963**	**1964**	**1964**	**1964**
Engine make	Villiers	Villiers	Villiers	Villiers
Engine type	A + Parkinson	36A or Starmaker	32A	32A + Parkinson
Gearbox type	4	4	4 wide	4
Front tyre (in.)		2·75 × 21	2·75 × 21	2·75 × 21
Rear tyre (in.)		4·00 × 18	4·00 × 18	4·00 × 18
Front brake dia. (in.)		6	6	6
Rear brake dia. (in.)		6	6	6
Front suspension	leading link	leading link	leading link	leading link
Rear suspension	s/a	s/a	s/a	s/a
Petrol tank (Imp. gal)	2·0	1·5	1·5 **1**	1·5 **1**
Wheelbase (in.)		52·5	51·5	51·5
Ground clear. (in.)			9·5	9·5
Seat height (in.)		30	28·5	28·5
Dry weight (lb)	210	230	223	215

1 option—2·25

Commander

Model	**I**	**II**	**III**
Year	**1952**	**1952**	**1952**
Engine make	Villiers	Villiers	Villiers
Engine type	2F	1F	10D
Gearbox type	1	2	3
Front tyre (in.)	2·25 × 21	2·50 × 19	3·00 × 19
Rear tyre (in.)	2·25 × 21	2·50 × 19	3·00 × 19
Front brake dia. (in.)	4	4	5
Rear brake dia. (in.)	back pedal 4	4	5
Front suspension	leading link	leading link	leading link
Rear suspension	s/a	s/a	s/a
Petrol tank (Imp. gal)	1·25	1·25	1·25
Wheelbase (in.)	53·5	53·5	53·5
Dry weight (lb)			175

Corgi

Model	**Mk I & II**	**Mk IV**
Year from	**1948**	**1952**
Year to	**1952 1**	**1954**
Engine make	Excelsior	Excelsior
Engine type	Spryt	Spryt
Gearbox type	1	2

151

O/A ratio: top	5·8	5·25
Front tyre (in.)	2·25 × 12·5	2·25 × 12·5
Rear tyre (in.)	2·25 × 12·5	2·25 × 12·5
Front brake dia. (in.)	4	4
Rear brake dia. (in.)	4	4
Front suspension	rigid	teles
Rear suspension	rigid	rigid
Petrol tank (Imp. gal)	1·25	1·25
Wheelbase (in.)	39	39
Ground clear. (in.)	4	4
Seat height (in.)	24·5 or 26·5	24·5 or 26·5
Dry weight (lb)	95	110

1 Mk I—1948

Cotton

Model	Vulcan	Cotanza	Vulcan	Cotanza 325	Trials	Villiers Twin 1
Year from	**1954**	**1955**	**1955**	**1955**	**1956**	**1956**
Year to	**1956**	**1960**	**1963**	**1960**	**1963**	**1963**
Engine make	Villiers	Anzani	Villiers	Anzani	Villiers	Villiers
Engine type	8E	242	9E	322	9E	2T
Gearbox type		4	3 or 4	4	4	4
O/A ratio: top			6·2	5·4	8·6	6·2
Front tyre (in.)	3·00 × 19	3·00 × 19	3·00 × 19	3·00 × 19	2·75 × 21	3·00 × 19
Rear tyre (in.)	3·00 × 19	3·00 × 19	3·00 × 19 **2**	3·25 × 19	4·00 × 19	3·25 × 19
Front brake dia. (in.)	6	6	6	6	6	6
Rear brake dia. (in.)	6	6	6	6	6	6
Front suspension	teles	teles **3**	teles **4**	teles **3**	teles **3**	teles **3**
Rear suspension	rigid	s/a	s/a	s/a	s/a	s/a
Petrol tank (Imp. gal)		2·75	2·75	2·75	1·5	2·75
Wheelbase (in.)	52		52		53	52
Ground clear. (in.)	5			6		
Seat height (in.)		31		30	31	31
Dry weight (lb)		255		255	250	270

1 1959—Herald **2** 1960—3·25 × 19 **3** 1959—leading link **4** 1958—leading link

Model	Messenger	Scrambler	Double Gloucester	Vulcan Sports	Trials 250	Continental
Year from	**1958**	**1959**	**1960**	**1960**	**1960**	**1961**
Year to	**1964**	**1962**	**1963**	**1968**	**1966**	**1968**
Engine make	Villiers	Villiers	Villiers	Villiers	Villiers	Villiers
Engine type	3T	33A **1**	2T	9E	32A **2**	2T **3**
Gearbox type	4	4	4	3 or 4	4	4
O/A ratio: top	6·2	9·3	6·2	6·2 **4**	8·6	6·2
Front tyre (in.)	3·00 × 21 **5**	3·00 × 21	3·00 × 19	3·00 × 19	2·75 × 21	3·00 × 19
Rear tyre (in.)	3·25 × 19	3·50 × 19	3·25 × 19	3·25 × 19	4·00 × 19 **6**	3·25 × 19
Front brake dia. (in.)	7	6	6	6	6	7 **7**
Rear brake dia. (in.)	7	6	6	6	6	6·25 **7**
Front suspension	leading link	leading link	leading link	leading link	leading link	leading link
Rear suspension	s/a	s/a	s/a	s/a	s/a	s/a
Petrol tank (Imp. gal)	2·75	1·5	2·75	2·75	1·5	2·75
Wheelbase (in.)	52	53	52	52	50·5	
Seat height (in.)	32	31	31	31	30·5	
Dry weight (lb)	280	245	260	245	245	280

1 1961—or 34A, 1962 only 34A **2** 1963—Special with 36A and Parkinson **3** 1965—4T **4** 1967—6·9 **5** 1963—3·00 × 19
6 1963—4·00 × 18 **7** 1963—Sports 6

Villiers Singles & Twins

Model	Corsair	Cougar	Cobra	Telstar	Conquest	Trials Starmaker
Year from	**1961**	**1961**	**1962**	**1962**	**1964**	**1964**
Year to	**1964**	**1966**	**1966**	**1966**	**1968**	**1968**
Engine make	Villiers	Villiers/Cross **1**	Villiers	Villiers	Villiers	Villiers
Engine type	31A	34A + alloy barrel	Starmaker	Starmaker	Starmaker	Starmaker
Gearbox type	4	4	4	4	4	4
O/A ratio: top	6·2	9·3	9·75	5·25	6·25	8·6
Front tyre (in.)	3·00 × 19	3·00 × 21	3·00 × 21	3·00 × 21 **2**	3·00 × 19	2·75 × 21
Rear tyre (in.)	3·25 × 19	4·00 × 18	4·00 × 18	3·50 × 18 **3**	3·25 × 19	4·00 × 18
Front brake dia. (in.)	7	6	6	7 **4**	7	6
Rear brake dia. (in.)	6·3	6	6	6	6	6
Front suspension	leading link	leading link	leading link	leading link	leading link	leading link
Rear suspension	s/a	s/a	s/a	s/a	s/a	s/a
Petrol tank (Imp. gal)	2·75	1·5	1·5 **5**	2·75	3·0	2·25
Dry weight (lb)	255	231	230	245	260	230

1 1963—Parkinson on 36A   **2** 1965—2·75 × 19   **3** 1965—3·25 × 19   **4** from 1964   **5** 1966—1·25

Model	Trials 37A	Telstar
Year from	**1966**	**1967**
Year to	**1968**	**1968**
Engine make	Villiers	Villiers
Engine type	37A	Starmaker
Gearbox type	4	4
O/A ratio: top	8·6	5·7
Front tyre (in.)	2·75 × 21	2·75 × 18
Rear tyre (in.)	4·00 × 18	3·25 × 18
Front suspsension	leading link	leading link
Rear suspension	s/a	s/a
Petrol tank (Imp. gal)	2·0	3·0
Dry weight (lb)	198	210

Cyc-Auto

Model	Autocycle	Carrier	Superior
Year from	**1946**	**1949**	**1949**
Year to	**1949**	**1958**	**1958**
Engine make	Cyc-Auto	Cyc-Auto	Cyc-Auto
Engine type	98 cc	98 cc	98 cc
Gearbox type	1	1	1
O/A ratio: top	11·0	11·0	11·0
Front tyre (in.)	26 × 2	26 × 2	26 × 2
Rear tyre (in.)	26 × 2	26 × 2	26 × 2
Front suspension	girder	strutted **1**	girder
Rear suspension	rigid	rigid	rigid
Petrol tank (Imp. gal)			1·5
Dry weight (lb)			124

1 1952 on—girder

DMW

Model	**125**	**200**	**4S Competition**	**Coronation**	**200 de luxe**	**4S Competition 1**
Year from	**1951**	**1951**	**1952**	**1953**	**1954**	**1954**
Year to	**1953**	**1953**	**1953**	**1953**	**1955**	**1956**
Engine make	Villiers	Villiers	Villiers	Villiers	Villiers	Villiers
Engine type	10D	6E	10D or 6E	10D	8E	7E
Gearbox type	3	3	3	3	3 or 4	3 or 4
O/A ratio: top	7·6	5·9	6·8	7·0	5·8	6·8
Front tyre (in.)	2·75 × 19	2·75 × 19 **2**	2·75 × 19	2·75 × 19	3·25 × 18	2·75 × 19
Rear tyre (in.)	3·00 × 19	3·00 × 19	3·25 × 19 **3**	3·00 × 19	3·25 × 18	4·00 × 19
Front brake dia. (in.)	5	5	5	5		
Rear brake dia. (in.)	5	5	5			
Front suspension	teles	teles	teles	bottom link	teles **4**	teles **4**
Rear suspension	plunger **5**	plunger **5**	plunger **6**	rigid	plunger	plunger
Petrol tank (Imp. gal)	2·5	2·5	2·5	2·5	2·5	2·5
Wheelbase (in.)			47	48	47	47
Ground clear. (in.)			7·5	7	7	8
Seat height (in.)	28·5	28·5	31	28·5–32	29	30
Dry weight (lb)	150	195	180	150	198	197

1 1955—5S **2** 1952—3·00 **3** 1953—4·00 for 197 **4** opt—Earles leading link **5** std—rigid **6** 1952 only—rigid, then as option

Model	**200P 1**	**Moto-Cross 2**	**Cortina**	**150 de luxe**	**150P Leda**	**200P Mk 9**
Year from	**1954**	**1954**	**1954**	**1955**	**1955**	**1956**
Year to	**1957**	**1956**	**1957**	**1955**	**1956**	**1964**
Engine make	Villiers	Villiers	Villiers	Villiers	Villiers	Villiers
Engine type	8E	7E	1H	29C	29C	9E
Gearbox type	3 or 4	3 or 4	4		4	4
O/A ratio: top	5·8		6·1		5·8	6·2
Front tyre (in.)	3·25 × 18		3·25 × 18	3·25 × 18	3·25 × 18	3·25 × 18
Rear tyre (in.)	3·25 × 18		3·25 × 18	3·25 × 18	3·25 × 18	3·25 × 18
Front brake dia. (in.)	5	5	6		6	6
Rear brake dia. (in.)	5	5	6		6	6
Front suspension	teles **3**	teles **4**	teles **3**	teles **3**	teles **3**	teles **3**
Rear suspension	s/a	s/a	s/a	plunger	s/a	s/a
Petrol tank (Imp. gal)	3·0	2·5	3·5	2·5	3·0	3·5
Wheelbase (in.)	50	50	51	47	50	50
Ground clear. (in.)	6	7	6	7	6	6
Seat height (in.)	29	29	29	29	29	29
Dry weight (lb)	230	218	253		224	244

1 1956—200P Mk 1 **2** 1955—Mk 5 M-X **3** opt—Earles leading link **4** 1955—leading link

Model	**Mk 6 M-X**	**Mk 7 Trials**	**150P Mk 9**	**175P Mk 9**	**Mk 8**	**Dolomite II**
Year from	**1956**	**1956**	**1956**	**1956**	**1956**	**1956**
Year to	**1957**	**1957**	**1957**	**1957**	**1958**	**1963**
Engine make	Villiers	Villiers	Villiers	Villiers	Villiers	Villiers
Engine type	9E	9E	31C	2L	8E	2T
Gearbox type	4	3 or 4	4	4	3	4
O/A ratio: top	6·9	6·8	7·0	6·6	5·8	6·1
Front tyre (in.)			3·25 × 18	3·25 × 18	3·00 × 19	3·25 × 18
Rear tyre (in.)		4·00 ×	3·25 × 18	3·25 × 18	3·00 × 19	3·25 × 18
Front brake dia. (in.)	6			6	5	6
Rear brake dia. (in.)	6			6	5	6
Front suspension	Earles	teles	teles	teles	teles	teles **1**
Rear suspension	s/a	s/a	s/a	s/a	s/a	s/a
Petrol tank (Imp. gal)	2·5	2·5	3·0	3·5	3·0	3·5

Villiers Singles & Twins

Wheelbase (in.)	50	50	50	50	50	51
Ground clear. (in.)	7	8	6	4·5	8	6
Seat height (in.)	29	30	29	29	30	29
Dry weight (lb)	220	244	240	244	210	253

1 opt—Earles

Model	Mk 10	Mk 12 Trials	Mk 12 M-X	Dolomite IIA	Mk 14 M-X	Mk 15 Trials
Year from	1957	1959	1959	1959	1961	1961
Year to	1961	1961	1961	1962	1962	1962
Engine make	Villiers	Villiers	Villiers	Villiers	Villiers	Villiers
Engine type	2T	32A **1**	33A **1**	3T	34A	32A
Gearbox type	4 close or wide	4	4	4	4	4
O/A ratio: top	optional	optional	optional	5·9	9·8	8·0
Front tyre (in.)	2·75 × 21	2·75 × 21	2·75 × 21	3·25 × 18	2·75 × 21	2·75 × 21
Rear tyre (in.)	4·00 × 18	4·00 × 18	3·50 × 19	3·25 × 18	4·00 × 18	4·00 × 18
Front brake dia. (in.)	6	6	6	6	6	6
Rear brake dia. (in.)	6	6	6	6	6	6
Front suspension	Earles	Earles	Earles	teles **2**	Earles	Earles
Rear suspension	s/a	s/a	s/a	s/a	s/a	s/a
Petrol tank (Imp. gal)	2·5	2·5	2·5	3·5	2·25	2·25
Wheelbase (in.)	50·5	50·5	50·5	52	52·5	52·5
Ground clear. (in.)						9·5
Seat height (in.)	31	31	31	30	33·5	33·5
Dry weight (lb)	288	268	268	300	256	256

1 opt—9E **2** opt—Earles

Model	Dolomite II Sports	Mk 16 M-X	Mk 17 Trials	Hornet (RR)	Mk 18	Mk 19
Year from	1962	1963	1963	1963	1963	1964
Year to	1963	1963	1965	1967	1965	1965
Engine make	Villiers	Villiers	Villiers	Villiers	Villiers	Villiers
Engine type	2T	Starmaker	32A	Starmaker	Starmaker	36A
Gearbox type	4	4	4	4	4	4
O/A ratio: top		optional	optional	optional	optional	optional
Front tyre (in.)	3·25 × 18	2·75 × 21	2·75 × 21	3·00 × 18	2·75 × 21	2·75 × 21
Rear tyre (in.)	3·25 × 18	4·00 × 18	4·00 × 18	3·00 × 18	4·00 × 18	4·00 × 18
Front brake dia. (in.)	6	6	6	6 twin	6	6
Rear brake dia. (in.)	6	6	6	6	6	6
Front suspension	teles	Earles	Earles	teles	Earles	Earles
Rear suspension	s/a	s/a	s/a	s/a	s/a	s/a
Petrol tank (Imp. gal)	3·5		2·0	4·0	2·0	2·0
Wheelbase (in.)	51					
Ground clear. (in.)	6					
Seat height (in.)	29					
Dry weight (lb)			256	205	250	

Model	Mk 14 M-X	Dolomite II	Sports Twin
Year from	1963	1964	1964
Year to	1964	1965	1966
Engine make	Villiers	Villiers	Villiers
Engine type	36A	4T	4T
Gearbox type	4	4	4
O/A ratio: top	9·75	6·2	6·2
Front tyre (in.)	2·75 × 21	3·25 × 18	3·25 × 18
Rear tyre (in.)	4·00 × 18	3·25 × 18	3·25 × 18
Front brake dia. (in.)	6	6	6

155

Part Three Appendix

Rear brake dia. (in.)	6	6	6
Front suspension	Earles	teles	teles
Rear suspension	s/a	s/a	s/a
Petrol tank (Imp. gal)	2·0	3·5	3·5
Wheelbase (in).		52	52
Ground clear. (in.)		5·75	5·75
Seat height (in.)		30	30
Dry weight (lb)	245	298	285

Dot

Model	125	200	Scrambles	Trials	Scramblers	Mancunian
Year from	1949	1949	1950	1954	1954	1955
Year to	—	1953	1953	1956	1956	1958
Engine make	Villiers	Villiers	Villiers	Villiers	Villiers	Villiers
Engine type	10D	6E	6E	8E	8E	9E
Gearbox type	3	3	3	3 or 4	3 or 4	4 **1**
O/A ratio: top	7·2	5·87 **2**	7·47 **3**	6·5	7·68	
Front tyre (in.)	2·75 × 19	3·00 × 19	3·00 × 19	2·75 × 21 or 3·00 × 19	3·00 × 19	3·00 × 19
Rear tyre (in.)	2·75 × 19	3·00 × 19	3·25 × 19	3·50 × 19	3·25 × 19	3·25 × 19
Front brake dia. (in.)		5	5	5 **4**	5 **4**	6 twin
Rear brake dia. (in.)		6	6	6	6	6
Front suspension	girder	girder **5**	teles	teles **6**	teles **6**	leading link
Rear suspension	rigid	rigid **7**	rigid **8**	rigid **9**	s/a	s/a
Petrol tank (Imp. gal)	2·75	2·75	1·5	2·5	2·5	2·75
Wheelbase (in.)		52·4	49·5	51·5	51·5	
Ground clear. (in.)			7	8	7	
Seat height (in.)		27	29·5	31	30	
Dry weight (lb)	180	200	190	220	220	

1 1958—or 3 **2** 6·2 prototype **3** trials—6·4 or 6·94 **4** 6 with Earles **5** 1951—teles **6** Earles option **7** 1953—also s/a **8** 1953—s/a **9** or s/a

Model	Trials	Scrambles	Trials	Scrambles	Twin scrambler	350 scrambler
Year from	1956	1956	1959	1959	1959	1959
Year to	1962	1960	1968	1962	1959	1960
Engine make	Villiers	Villiers	Villiers	Villiers	Villiers	RCA
Engine type	9E	9E	32A **1**	31A **2**	2T	349
Gearbox type	3 or 4	3 or 4	3 or 4	3 or 4	4	4
Front tyre (in.)	2·75 × 21	2·75 × 21	2·75 × 21	2·75 × 21	2·75 × 21	2·75 × 21
Rear tyre (in.)	3·25 × 19 **3**	3·25 × 19	3·50 × 19 **3**	3·50 × 19	3·50 × 19	4·00 × 18
Front brake dia. (in.)	6	6	6	6	6	6
Rear brake dia. (in.)	6	6	6	6	6	6
Front suspension	leading link	leading link	leading link	leading link	leading link	leading link
Rear suspension	s/a	s/a	s/a	s/a	s/a	s/a
Petrol tank (Imp. gal)	2·75	2·75	1·75 **4**	1·75	1·75	1·75
Wheelbase (in.)	51·5	52	52 **5**	52	52	52
Ground clear. (in.)	9					
Seat height (in.)	30					
Dry weight (lb)	235	230	234	232	265	275

1 1967—37A **2** 1961—34A **3** Works replica—4·00 × 18 **4** 1962—2·0 **5** 1962—53·2

Model	350 road	Demon	WR Trials	Demon International	360 Demon
Year from	1959	1962	1962	1962	1967
Year to	1960	1968	1968		1968
Engine make	RCA	Villiers **1**	Villiers	Villiers	Alpha Dot

Engine type	349	34A **2**	32A	Starmaker	360	
Gearbox type	4	4	4	4	4	
Front tyre (in.)	3·00 × 19	2·75 × 21	2·75 × 21	2·75 × 21	2·75 × 21	
Rear tyre (in.)	3·25 × 18	4·00 × 18	4·00 × 18	4·00 × 18	4·00 × 18	
Front brake dia. (in.)	6	6	7	6	6	
Rear brake dia. (in.)	6	6	6	6	6	
Front suspension	leading link	leading link	leading link	leading link	leading link	
Rear suspension	s/a	s/a	s/a	s/a	s/a	
Petrol tank (Imp. gal)	1·75 or 3	2·0	2·0			
Wheelbase (in.)		54	53·5	54		
Ground clear. (in.)		9·5	9·5	9·5		
Seat height (in.)		31	31	31		
Dry weight (lb)	280	220	225	220		

1 1965—Alpha/Dot **2** 1963—36A

Excelsior

Model	Autobyk	V1	G2	S1	O	Minor M1
Year from	**1946**	**1947**	**1947**	**1947**	**1946**	**1948**
Year to	**1946**	**1949**	**1956**	**1956**	**1948**	**1949**
Engine make	Villiers	Villiers	Excelsior	Excelsior	Villiers	Excelsior
Engine type	JDL	JDL	Goblin	Spryt	9D	Goblin 98
Gearbox type	1	1	2	1	3	2
O/A ratio: top		8·5		11·3	8·1	
Front tyre (in.)		26 × 2	26 × 2 **1**	26 × 2	2·75 × 19	2·50 × 19
Rear tyre (in.)		26 × 2	26 × 2 **1**	26 × 2	2·75 × 19	2·50 × 19
Front brake dia. (in.)		4	4	4	4	
Rear brake dia. (in.)		4	4	4	4	
Front suspension	rigid	girder	girder	girder	girder	girder
Rear suspension	rigid	rigid	rigid	rigid	rigid	rigid
Petrol tank (Imp. gal)		1·4	1·4	1·4	2·75	1·7
Wheelbase (in.)			50	50		
Ground clear. (in.)			5	5		
Seat height (in.)			32	32		
Dry weight (lb)			130	121	145	135

1 by 1951—2·25 × 21

Model	Minor M2	F4	F4S/F6S	SB1	F series **1**	C series **2**
Year from	**1948**	**1953**	**1956**	**1956**	**1959**	**1957**
Year to	**1949**	**1957**	**1958 3**	**1959**	**1962**	**1964**
Engine make	Excelsior	Villiers	Villiers	Villiers	Villiers	Villiers
Engine type	Goblin 123	4F **4**	6F	6F	6F	6F
Gearbox type	2	2	2	2	2	2
O/A ratio: top		7·48	8·5	8·5	8·5	8·5
Front tyre (in.)	2·50 × 19	2·25 × 19	2·25 × 19	2·25 × 19	2·25 × 19	2·25 × 19
Rear tyre (in.)	2·50 × 19	2·25 × 19	2·25 × 19	2·25 × 19	2·25 × 19	2·25 × 19
Front brake dia. (in.)		4	4 **5**	4·6	4	4
Rear brake dia. (in.)		4·6	4·6	4·6	4·6	4·6
Front suspension	girder	girder	girder	teles	teles	teles
Rear suspension	rigid	rigid	plunger	plunger	rigid	s/a
Petrol tank (Imp. gal)	1·7	1·75	1·75	1·75	1·75	1·75
Wheelbase (in.)		47	47	47		47
Ground clear. (in.)		6·5	5	5		4·2
Seat height (in.)		26	27	27		27
Dry weight (lb)	135	126	138	165	126	136

1 1959—F4F, 1960—F10, 1961—F11, 1962—F12 **2** 1958—CA8, 1959—CA9, 1960—C10, 1961—C11, 1962—C12, 1963/64—C14
3 F4S—1956 **4** 1956—6F **5** F6S—4·6

Villiers Singles & Twins

157

Part Three Appendix

Model	U1 & U2	D12	C series 1	C series 2	U series 3	U series 4
Year from	1949	1953	1952	1955	1959	1962
Year to	1954	1954	1957	1959	1961	1965
Engine make	Villiers	Villiers	Excelsior	Villiers	Villiers	Excelsior 5
Engine type	10D	12D or 13D	150	30C	31C	147 5
Gearbox type	3	3	3	3	3	3
O/A ratio: top		7·0	6·52	6·3 6	6·6	
Front tyre (in.)	2·75 × 19	2·25 × 19	3·00 × 19 7	2·50 × 19 8	3·00 × 19	2·75 × 19
Rear tyre (in.)	2·75 × 19 9	2·25 × 19	3·00 × 19 7	2·50 × 19 8	3·00 × 19	2·75 × 19
Front brake dia. (in.)	5	4	5	5	5	5
Rear brake dia. (in.)	5	4·6	5	5	5	5
Front suspension	teles	girder	teles	teles	teles	teles
Rear suspension	rigid 10	rigid	plunger 11	plunger 12	s/a	s/a
Petrol tank (Imp. gal)	2·75	2·0	2·75	2·0 13	2·5	2·5
Wheelbase (in.)			49·5 14	47 15		
Ground clear. (in.)			5·5 16	5 17		
Seat height (in.)			30 18	28 19		
Dry weight (lb)		146	222 20	170 21	195	224

1 C1—1953/54, C2—1952/54, C3—1954/56, C4—1956/57 **2** C1—1955/56, U8—1957/58, U9—1958/59 **3** U10—1959/60, U11—1960/61 **4** U12—1962, U14—1963/65 **5** 1963—Villiers 31C **6** 1957—6·7 **7** 1956 only—2·75 × 19 **8** 1957—2·75 × 19 **9** 1952—3·00 × 19 **10** 1950—plunger **11** 1954—s/a **12** 1957—s/a **13** 1957—2·75 **14** 1954—50 **15** 1957—50 **16** 1956—6 **17** 1957—6 **18** 1954—31 **19** 1957—31 **20** from 1954, 1956—200 **21** 1957—181

Model	R1 & R2	R series 1	A9	R10/11 2	Talisman Twin 3	S Twin 4
Year from	1949	1953	1955	1959	1949	1957
Year to	1954	1957	1957	1961	1962	1962
Engine make	Villiers	Villiers	Villiers	Villiers	Excelsior	Excelsior
Engine type	6E 5	8E	9E	9E	243 twin	328 twin
Gearbox type	3	3 or 4	4	4	4	4
O/A ratio: top		5·74 6	5·8	6·1	5·54 7	5·2 8
Front tyre (in.)	2·75 × 19 9	3·00 × 19	3·00 × 19	3·00 × 19	3·00 × 19	3·00 × 19 10
Rear tyre (in.)	2·75 × 19 9	3·00 × 19	3·00 × 19	3·00 × 19	3·00 × 19 11	3·25 × 19 10
Front brake dia. (in.)	5	5	5	5	5 12	6 13
Rear brake dia. (in.)	5	5	5	5	6	6
Front suspension	teles	teles	teles	teles	teles	teles
Rear suspension	rigid 14	s/a	s/a	s/a	plunger 15	s/a
Petrol tank (Imp. gal)	2·75	2·5 16	2·75	3·0	2·75 17	3·0 18
Wheelbase (in.)		53	53		49 19	53
Ground clear. (in.)		6	6		5 20	6
Seat height (in.)		33 21	32		29 22	32
Dry weight (lb)		200 23	204	205	220 24	296

1 R3—1953/54, R4—1953/55, R5—1955, R6—1955/57 **2** R10—1959/60, R11—1960/61 **3** TT1—1949/54, STT1—1952/54, TT2 and STT2—1954/55, TT3—1955/57, STT4—1955/56, STT5—1956/57, STT6—1957/58, TT4—1957/62 **4** S8—1957/59, S9—1959/60, S10—1960/61, ETT9—1961/62 **5** 1954—8E **6** R5, R6—6·3 **7** STT1—6·09 **8** not S8 **9** 1952—3·00 × 19 **10** S9 on—3·25 × 18 **11** STT6—3·25 × 19 **12** 1955—SE, STT4, STT5, STT6, TT4—6 **13** S9 on—7 **14** 1950—plunger **15** TT1, STT1 only **16** R5, R6—2·75 **17** 3·0 except TT1, STT1 **18** S9 on—3·5 **19** 53 except TT1, STT1 **20** 6 except TT1, STT1 **21** R5, R6—32 **22** TT models 30, STT—32 except TT1, STT1 **23** R5, R6—185 **24** 1952 TT1—237, STT1—265, TT2—268, TT3—230, STT4—240, STT6—290, TT4—270

FLM

Model	Glideride
Year from	1951
Year to	1953
Engine make	JAP
Engine type	125

158

Gearbox type	3
O/A ratio: top	7·2
Front tyre (in.)	2·75 × 19
Rear tyre (in.)	2·75 × 19
Front brake dia. (in.)	5
Rear brake dia. (in.)	5
Front suspension	teles
Rear suspension	s/a
Petrol tank (Imp. gal)	3.0

Villiers Singles & Twins

Francis-Barnett

Model	50 Powerbike	51 Merlin	52/53 Merlin	54/55 Falcon	56 Powerbike	57 Merlin
Year from	**1945**	**1946**	**1948**	**1948**	**1949**	**1951**
Year to	**1949**	**1948**	**1953**	**1953**	**1952**	**1953**
Use	road	road	road	road	road	road
Engine make	Villiers	Villiers	Villiers	Villiers	Villiers	Villiers
Engine type	JDL	9D	10D **1**	6E **2**	2F	10D **1**
Gearbox type	1	3	3	3	1	3
O/A ratio: top	11·8	8·1	7·08 **3**	5·6 **4**	11	
Front tyre (in.)	26 × 2	3·00 × 19	3·00 × 19	3·00 × 19	2·25 × 21	3·00 × 19
Rear tyre (in.)	26 × 2	3·00 × 19	3·00 × 19	3·00 × 19	2·25 × 21	3·00 × 19
Front brake dia. (in.)	3·62	5	5	5	4	5
Rear brake dia. (in.)	4	5	5	5	4	5
Front suspension	girder	girder	teles	teles	girder	teles
Rear suspension	rigid	rigid	rigid	rigid	rigid	s/a
Petrol tank (Imp. gal)	1·75	2·25	2·25	2·25	1·5	2·25
Wheelbase (in.)		50·5	49	49		49·5
Ground clear. (in.)		5	5	5		5
Seat height (in.)		26	27·75—29	27·75—29		28—29
Dry weight (lb)		178	181/193	187/199		212

1 1953—12D **2** 1953—8E **3** late 1949—7·18 **4** late 1949—5·87

Model	58 Falcon	59 Merlin	60 Falcon	61 Merlin	62 Falcon	63 Merlin
Year from	**1951**	**1951**	**1951**	**1953**	**1953**	**1953**
Year to	**1953**	**1952**	**1952**	**1953**	**1955**	**1953**
Use	road	comp	trials	trials	trials	M-X
Engine make	Villiers	Villiers	Villiers	Villiers	Villiers	Villiers
Engine type	6E **1**	10D	6E	10D	7E	10D
Gearbox type	3	3	3	4	4	4
O/A ratio: top	5·87 **2**	9·1	6·8		6·25	
Front tyre (in.)	3·00 × 19	2·75 × 21	2·75 × 21	2·75 × 21	2·75 × 21	2·75 × 21
Rear tyre (in.)	3·00 × 19	3·25 × 19	3·50 × 19	3·25 × 19	3·50 × 19	3·50 × 19
Front brake dia. (in.)	5	5	5	5	5	5
Rear brake dia. (in.)	5	5	5	5	5	5
Front suspension	teles	teles	teles	teles	teles	teles
Rear suspension	s/a	rigid	rigid	rigid	rigid	s/a
Petrol tank (Imp. gal)	2·25	2·25	2·25	2·25	2·25	2·25
Wheelbase (in.)	49·5	49	49	49·5	49·5	50·2
Ground clear. (in.)	5	7	7	7·5	7·5	7·5
Seat height (in.)	28—29	32·25—33·5	32·25—33·5	30·5—31·5	30·5—31·5	31—32
Dry weight (lb)	218		178		200	

1 1953—8E **2** 1953—5·74

159

Part Three Appendix

Model	64 Falcon	65 Falcon	66 Kestrel	67 Falcon	68 Cruiser	69 Kestrel
Year from	**1953**	**1953**	**1953**	**1953**	**1953**	**1954**
Year to	**1954**	**1954**	**1954**	**1954**	**1954**	**1955**
Use	M-X	trail	road	road	road	road
Engine make	Villiers	Villiers	Villiers	Villiers	Villiers	Villiers
Engine type	7E	8E	13D	8E	1H	30C
Gearbox type	4	3	3	3	4	3
O/A ratio: top	6·25	6·25	7·4		6·23	7·4
Front tyre (in.)	2·75 × 21	2·75 × 21	2·75 × 19	3·00 × 19	3·00 × 19	2·75 × 19
Rear tyre (in.)	3·50 × 19	3·25 × 19	2·75 × 19	3·00 × 19	3·25 × 19	2·75 × 19
Front brake dia. (in.)	5	5	4	5	6	4
Rear brake dia. (in.)	5	5	5	5	6	5
Front suspension	teles	teles	teles	teles	teles	teles
Rear suspension	s/a	s/a	plunger	s/a	s/a	plunger
Petrol tank (Imp. gal)	2·25	2·25	2·25	2·75	3·5	2·25
Wheelbase (in.)	50·2	50·2	49	49·5	51	49
Ground clear. (in.)	7·5	5·7	5	5	6·5	5
Seat height (in.)	31—32	31—32	28	28—29	31	28
Dry weight (lb)	212	221	164		280	170

Model	70 Falcon	71 Cruiser	72 Falcon	73 Plover	74 Falcon	75 Cruiser
Year from	**1954**	**1954**	**1954**	**1955**	**1955**	**1955**
Year to	**1955**	**1955**	**1955**	**1956**	**1957**	**1957**
Use	road	road	M-X	road	road	road
Engine make	Villiers	Villiers	Villiers	Villiers	Villiers	Villiers
Engine type	8E	1H	7E	30C	8E	1H
Gearbox type	3 or 4	4	4	3	3 or 4	4
O/A ratio: top	5·75	6·23	6·25	6·5	6·27	6·23
Front tyre (in.)	3·00 × 19	3·00 × 19	2·75 × 21	3·00 × 18	3·25 × 18	3·25 × 18
Rear tyre (in.)	3·00 × 19	3·25 × 19	3·50 × 19	3·00 × 18	3·25 × 18	3·25 × 19
Front brake dia. (in.)	5	6	5	4	5	6
Rear brake dia. (in.)	5	6	5	5	5	6
Front suspension	teles	teles	teles	teles	teles	teles
Rear suspension	s/a	s/a	s/a	s/a	s/a	s/a
Petrol tank (Imp. gal)	2·75	3·5	2·25	2·25	2·75	3·5
Wheelbase (in.)	49·8	51	50·2	49·5	49·8	51
Ground clear. (in.)	6	6·5	7·5	6	6	6·5
Seat height (in.)	31	31·5	31	29·5	31	31·5
Dry weight (lb)	243	280	212	190	243	280

Model	76 Falcon	77 Falcon	78 Plover	79 Light Cruiser	80 Cruiser	81 Falcon
Year from	**1955**	**1955**	**1956**	**1958**	**1956**	**1957**
Year to	**1957**	**1957**	**1959**	**1960**	**1963**	**1959**
Use	trials	M-X	road	road	road	road
Engine make	Villiers	Villiers	Villiers	AMC	AMC	Villiers
Engine type	7E	7E	30C	175	250	10E
Gearbox type	4	4	3	4	4	3
O/A ratio: top	6·5	6·5	6·5 **1**	6·9	5·9	6·1
Front tyre (in.)	2·75 × 21	2·75 × 21	3·00 × 18	3·00 × 18	3·25 × 18	3·25 × 18
Rear tyre (in.)	4·00 × 18	3·50 × 19	3·00 × 18	3·00 × 18	3·25 × 18	3·25 × 18
Front brake dia. (in.)	5	5	4	5	6	5
Rear brake dia. (in.)	5	5	5	5	6	5
Front suspension	teles	teles	teles	teles	teles	teles
Rear suspension	s/a	s/a	s/a	s/a	s/a	s/a
Petrol tank (Imp. gal)	2·0	2·0	2·25	3·75	3·75	2·75

160

Wheelbase (in.)	51·5	51·5	49·5	51	51·5	49·7
Ground clear. (in.)	8·2	8·2	6	5·5	5·5	6
Seat height (in.)	32	31·5	29·5	29·5	30	30·5
Dry weight (lb)	230	233	190	247	290	244

1 1958—6·64

Model	82 Scrambler	83 Trials	84 Cruiser	85 Trials	86 Plover	87 Falcon
Year from	1958	1958	1959	1959	1959	1959
Year to	1962	1959	1962	1962	1962	1966
Use	M-X	trials	road	trials	road	road
Engine make	AMC	AMC	AMC	AMC	AMC	AMC
Engine type	250	250	250	250	150	200
Gearbox type	4 close	4 wide	4	4 wide	3	4
O/A ratio: top	8·03	8·0	5·9	8·0	6·9	6·5
Front tyre (in.)	2·75 × 21	2·75 × 21	3·25 × 18	2·75 × 21	3·00 × 18	3·25 × 18
Rear tyre (in.)	3·50 × 19	4·00 × 19	3·25 × 18	4·00 × 19	3·00 × 18	3·25 × 18
Front brake dia. (in.)	6	6	6	5	4	5
Rear brake dia. (in.)	5	5	6	5	5	5
Front suspension	teles	teles	teles	teles	teles	teles
Rear suspension	s/a	s/a	s/a	s/a	s/a	s/a
Petrol tank (Imp. gal)	2·0	2·0	3·75	2·25	2·25	3·25
Wheelbase (in.)	53	53	52·2	52·5		
Ground clear. (in.)	8·5	8·5	6	8		
Seat height (in.)	31	34	30	31		
Dry weight (lb)	263	273	307	271	171	268

Model	88 Fulmar	89 Cruiser Twin	90 Fulmar Sports	91 Cruiser Twin Sports	92 Trials	93 Scrambler
Year from	**1961**	**1961**	**1962**	**1962**	**1962**	**1962**
Year to	**1965**	**1966**	**1965**	**1966**	**1966**	**1964**
Use	road	road	road	road	trials	M-X
Engine make	AMC	Villiers	AMC	Villiers	Villiers	Villiers
Engine type	150	2T **1**	150	2T **1**	32A	36A + Parkinson
Gearbox type	3	4	4	4	4 wide	4 close
O/A ratio: top	6·9	6·2	6·9	6·2	6·9	8·6
Front tyre (in.)	3·00 × 18	3·25 × 18	3·00 × 18	2·75 × 19	2·75 × 21	2·75 × 21
Rear tyre (in.)	3·00 × 18	3·25 × 18	3·00 × 18	3·25 × 18	4·00 × 19	4·00 × 18
Front brake dia. (in.)	5	6	5		5	6
Rear brake dia. (in.)	5	6	5		5	5
Front suspension	leading link	teles	leading link	teles	teles	teles
Rear suspension	s/a	s/a	s/a	s/a	s/a	s/a
Petrol tank (Imp. gal)	2·25	3·75	2·25	2·75	2·25	1·5
Wheelbase (in.)	49·5					
Ground clear. (in.)	5					
Seat height (in.)	29					
Dry weight (lb)	225	299	223	295	246	235

1 1964—4T

Model	94 Starmaker Scrambler	95 Plover	Model 96
Year from	**1963**	**1963**	**1965**
Year to	**1966**	**1965**	**1966**
Use	M-X	road	road
Engine make	Villiers	AMC	AMC
Engine type	Starmaker	150	150
Gearbox type	4	3	3

Villiers Singles & Twins

161

O/A ratio: top	10·75	6·7	6·7		
Front tyre (in.)	2·75 × 21	3·00 × 18	3·00 × 18		
Rear tyre (in.)	4·00 × 18	3·00 × 18	3·00 × 18		
Front brake dia. (in.)	6	5	5		
Rear brake dia. (in.)	5	5	5		
Front suspension	teles	teles	teles		
Rear suspension	s/a	s/a	s/a		
Petrol tank (Imp. gal)	1·5	2·75	2·25		
Dry weight (lb)	251	200	165		

Greeves

Model	20 Road	20S	20T	25D & R	32D	20D
Year from	**1953**	**1953**	**1953**	**1953**	**1954**	**1955**
Year to	**1957**	**1955**	**1955**	**1957**	**1957**	**1958**
Engine make	Villiers	Villiers	Villiers	Anzani	Anzani	Villiers
Engine type	8E **1**	8E	8E	242	322	9E
Gearbox ratio	3 or 4	4 close	4 wide	4	4	4
O/A ratio: top	6·26	7·3	6·8	5·8 **2**	5·38	6·1
Front tyre (in.)	2/75 or 3·00 × 19	3·00 × 19	2·75 × 21	3·00 × 19	3·00 × 19 **3**	2·75 × 19
Rear tyre (in.)	3·25 × 19	3·25 × 19	3·50 × 19 **4**	3·25 × 19	3·25 × 19	3·25 × 19
Front brake dia. (in.)	6 **5**	6	6	6	twin 6	6
Rear brake dia. (in.)	6 **5**	6	6	6	7	6
Front suspension	leading link	leading link	leading link	leading link	leading link	leading link
Rear suspension	s/a	s/a	s/a	s/a	s/a	s/a
Petrol tank (Imp. gal)	2·5	2·5	2·5	2·5	2·5	3·6
Wheelbase (in.)	52	52	52	52	52	52
Ground clear. (in.)	7	7·8	8·2	7	7	6·7
Seat height (in.)	32	32·5	33	31·5	32	31
Dry weight (lb)	240	213	225	252	276	248

1 1957—7E **2** 1955—6·1 **3** 1957—3·00 × 20 **4** 1954—4·00 × 18 **5** 1954 20R—5

Model	20S	20T	25D	20SA-SC	20TA-TE	25 Sports Twin
Year from	**1955**	**1955**	**1956**	**1957**	**1957**	**1958**
Year to	**1957**	**1957**	**1958**	**1962**	**1965**	**1966**
Engine make	Villiers	Villiers	Villiers	Villiers	Villiers	Villiers
Engine type	9E	9E	2T	9E	9E	2T **1**
Gearbox ratio	4	4	4	4	4	4
O/A ratio: top	7·1	6·3	6·2	8·6	7·8	6·2
Front tyre (in.)	2·75 × 21 **2**	2·75 × 21 **2**	3·00 × 20	2·75 × 21	2·75 × 21	2·75 × 19
Rear tyre (in.)	3·50 × 19	4·00 × 18	3·25 × 19	3·25 × 19	4·00 × 18	3·25 × 18
Front brake dia. (in.)	6	6	twin 6	6	6	6
Rear brake dia. (in.)	6	6	6	6	6	6
Front suspension	leading link	leading link	leading link	leading link	leading link	leading link
Rear suspension	s/a	s/a	s/a	s/a	s/a	s/a
Petrol tank (Imp. gal)	2·5	2·5	3·6	1·75 **3**	1·75 **3**	2·5
Wheelbase (in.)	52	52	52	52	52 **4**	52
Ground clear. (in.)	8	8·5	6·7	8	9·5	6·5
Seat height (in.)	31	31·5	31	31	31·5	31·5
Dry weight (lb)	218	228	294	218	228	270

1 1964—4T **2** 1957—3·00 × 20 **3** 1960—2·0 **4** 1960—51·5

Villiers Singles & Twins

Model	24T	24S	24D	32D	20DC	24TE-TJ
Year from	1958	1958	1958	1960	1961	1961
Year to	1961	1964	1962	1963	1966	1968
Engine make	Villiers	Villiers	Villiers	Villiers	Villiers	Villiers
Engine type	31A [1]	31A [2]	31A [1]	3T	9E	32A [3]
Gearbox type	4	4	4	4	4	4
O/A ratio: top	7·8	8·6 [4]	5·7	5·7	6·6	7·8 [5]
Front tyre (in.)	2·75 × 21	2·75 × 21	2·75 × 19	2·75 × 19	2·75 × 19	2·75 × 21
Rear tyre (in.)	4·00 × 18	3·50 × 19 [6]	3·25 × 18	3·25 × 18	3·25 × 18	4·00 × 18 [7]
Front brake dia. (in.)	6	6	6	6	6	6
Rear brake dia. (in.)	6	6	6	6	6	6
Front suspension	leading link	leading link	leading link	leading link	leading link	leading link [8]
Rear suspension	s/a	s/a	s/a	s/a	s/a	s/a
Petrol tank (Imp. gal)	1·75 [9]	1·75 [9]	2·5	2·5	2·5	1·75
Wheelbase (in.)	52 [10]	52	52	52	52	52
Ground clear. (in.)	9·5	8		6·5		9
Seat height (in.)	31·5	31	31·5	31·5		32
Dry weight (lb)	240	230	245	270	238	235 [11]

[1] 1960—32A [2] 1960—34A, 1963—36A [3] 1965—TFS-Challenger top half, 1967—37A bottom half [4] 1963—10·0 [5] 1965—8·3
[6] 1962—4·00 × 18 [7] 1966—4·00 × 19 [8] 1968—teles [9] 1960—2·0 [10] 1960—51·5 [11] 1967—215

Model	24RAS	24ME	24MX	24RBS-RES	36MX	35RFS
Year from	1962	1963	1964	1964	1966	1968
Year to	1963	1964	1967	1968	1968	1968
Engine make	Villiers	Villiers	Greeves	Greeves	Greeves	Greeves
Engine type	36A/Greeves	Starmaker	Challenger	Challenger	360	344
Gearbox type	4	4	4	5	4	5
O/A ratio: top	5·2 to 6·0		10·4		10·3	
Front tyre (in.)	2·75 × 19 [1]	2·75 × 21	2·75 × 21	2·75 × 18	2·75 × 21	2·75 × 18
Rear tyre (in.)	3·25 × 18 [1]	4·00 × 18	4·00 × 18	2·75 × 18	4·00 × 18	2·75 × 18
Front brake dia. (in.)	6	6	6	6 [2]	6	7
Rear brake dia. (in.)	6	6	6	6	6	6
Front suspension	leading link	leading link	leading link	leading link	leading link [3]	leading link
Rear suspension	s/a	s/a	s/a	s/a	s/a	s/a
Petrol tank (Imp. gal)	2·0	2·0	2·0	3·0	1·75	3·0
Wheelbase (in.)	52	52				
Dry weight (lb)	188		217	196	226	

[1] 1963—2·75 × 18, 3·00 × 18 [2] 1967—7 [3] teles option

Model	24MX4	250 Griffon	380 Griffon	Pathfinder
Year from	1967	1969	1969	1970
Year to	1968	1978	1978	1978
Engine make	Greeves	Greeves	Greeves	Puch
Engine type	246	246	380	169
Gearbox type	4	4	4	6
O/A ratio: top		10·7	9·0	10·1
Front tyre (in.)	2·75 × 21	3·00 × 21	3·00 × 21	2·75 × 21
Rear tyre (in.)	4·00 × 18	4·00 × 18	4·00 × 18	4·00 × 18
Front brake dia. (in.)	6	6	6	5
Rear brake dia. (in.)	6	6	6	5
Front suspension	leading link [1]	teles	teles	teles
Rear suspension	s/a	s/a	s/a	s/a
Petrol tank (Imp. gal)	1·75	1·5	1·5	1·25
Wheelbase (in.)		55	55	51

163

Ground clear. (in.)		10		10		11
Seat height (in.)		32·5		32·5		31
Dry weight (lb)		216		227		190

1 teles option

HJH

Model	Dragon	Super Dragon	Sports Dragon	Super Sports Dragon	Trials	Scrambler
Year from	**1954**	**1955**	**1955**	**1955**	**1955**	**1955**
Year to	**1956**	**1955**	**1956**	**1956**	**1956**	**1956**
Engine make	Villiers	Villiers	Villiers	Villiers	Villiers	Villiers
Engine type	8E	8E	8E	8E	7E	7E
Gearbox type	3	3	3	3 or 4	4	4
O/A ratio: top	5·8	5·8	5·8	5·8	6·5	7·7
Front tyre (in.)	3·00 × 19	3·00 × 19	3·00 × 19	3·00 × 19		
Rear tyre (in.)	3·00 × 19	3·00 × 19	3·00 × 19	3·00 × 19		
Front brake dia. (in.)	5	5	5	5	5	5
Rear brake dia. (in.)	5	5	5	5	5	5
Front suspension	teles	leading link	teles	leading link	leading link	leading link
Rear suspension	plunger **1**	plunger	s/a	s/a	rigid **2**	s/a
Petrol tank (Imp. gal)	2·5	2·5	2·5	2·75	1·25	1·25
Wheelbase (in.)	47 **3**	47	47	47·5	46	47
Ground clear. (in.)	9·5	9·5	9·5	9	10·5	9·5
Seat height (in.)	29	29	29	29	30	29
Dry weight (lb)	182		204	206	190	196

1 1956—rigid   **2** 1956—s/a   **3** 1956—46

Model	Dragonette	Sports Dragonette	Dragon Major
Year from	**1955**	**1955**	**1955**
Year to	**1956**	**1956**	**1956**
Engine make	Villiers	Villiers	Villiers
Engine type	30C	30C	1H
Gearbox type	3	3	4
O/A ratio: top	6·8	6·8	
Front suspension	teles	teles	leading link
Rear suspension	rigid	s/a	s/a
Petrol tank (Imp. gal)	2·5	2·5	3·0
Wheelbase (in.)	46	46	
Ground clear. (in.)	9·5	9·5	
Seat height (in.)	27·5	28	
Dry weight (lb)	175	185	

James

Model	Superlux	ML	Comet 1	Cadet	Captain	Superlux
Year from	**1946**	**1946**	**1949**	**1949**	**1948**	**1948**
Year to	**1949**	**1948**	**1955**	**1953**	**1953**	**1954**
Engine make	Villiers	Villiers	Villiers	Villiers	Villiers	Villiers
Engine type	JDL	9D	1F **2**	10D	6E	2F
Gearbox type	1	3	2	3	3	1
O/A ratio: top	12	8·1	8·47	7·93 **3**	6·4 **4**	10·78
Front tyre (in.)	26 × 2	2·75 × 19	2·50 × 19 **5**	3·00 × 19	3·00 × 19	2·25 × 21
Rear tyre (in.)	26 × 2	2·75 × 19	2·50 × 19 **5**	3·00 × 19	3·00 × 19	2·25 × 21
Front brake dia. (in.)	4	4	4	5	5	4

Villiers Singles & Twins

Rear brake dia. (in.)	3·6	5	4	5	5	4
Front suspension	girder	girder	girder	girder **6**	girder **6**	girder
Rear suspension	rigid	rigid	rigid	rigid **7**	rigid **8**	rigid
Petrol tank (Imp. gal)	1·4	2·25	1·75 **9**	2·25	2·25	1·5
Wheelbase (in.)		46·5	46·5	49	49	
Ground clear. (in.)			4·7	6·2	5	
Seat height (in.)	32	28	28·5	28·5	31	
Dry weight (lb)	100	160	128	168	190	134

1 1954—J11 **2** 1953—4F **3** 1950—7·185 **4** 1950—5·86 **5** std-2.25 x 21, 1954-2.25 x 19 **6** 1950—teles **7** 1952—plunger **8** 1950—plunger option, 1952—std **9** 1953—2·0

Model	J5 Cadet	J9 Commando	K7 Captain	K7C Cotswold	K12 Colonel	J15 Cadet
Year from	1953	1953	1953	1953	1953	1954
Year to	1954	1955	1959	1957	1957	1955
Engine make	Villiers	Villiers	Villiers	Villiers	Villiers	Villiers
Engine type	13D	6E **1**	8E **2**	7E	1H	30C
Gearbox type	3	3 **3**	3 **4**	4 close	4	3
O/A ratio: top	7·33	6·8 **5**	5·74 **6**	6·27	6·21	6·5
Front tyre (in.)	2·75 x 19	2·75 x 21	3·00 x 19 **7**	2·75 x 21	3·00 x 19 **8**	2·75 x 19
Rear tyre (in.)	2·75 x 19	4·00 x 19	3·00 x 19 **9 7**	3·50 x 19	3·00 x 19 **8**	2·75 x 19
Front brake dia. (in.)	4	5	5	5	6	4
Rear brake dia. (in.)	4 **10**	5	5	5	6	5
Front suspension	teles	teles	teles	teles	teles	teles
Rear suspension	plunger	rigid	s/a	s/a	s/a	plunger
Petrol tank (Imp. gal)	2·0	2·25 **11**	2·25 **11**	2·25 **11**	2·25 **11**	2·0
Wheelbase (in.)	49	49	50	50	50	49
Ground clear. (in.)	5	8·5	5·5	6·5	5·5	5
Seat height (in.)	28	30	30	30	30	28
Dry weight (lb)	166	196	220	208	275	170

1 1954—7E **2** 1958—10E **3** 1954—4 wide **4** 1955/56—4 option **5** 1954—6·54 **6** 1958—6·3 **7** 1956—3·00 x 18 **8** 1956—3·25 x 18 **9** 1957—3·25 x 18 **10** 1954—5 **11** 1955—2·75

Model	L1 Comet 100	L15 Cadet	K7T Commando	L25 Commodore	L17 Cavalier	K7T Commando
Year from	1955	1955	1955	1955	1957	1958
Year to	1966	1959	1957	1962	1959	1958
Engine make	Villiers	Villiers	Villiers	AMC	AMC	Villiers
Engine type	4F **1**	30C	7E	250	175	10E
Gearbox type	2	3	4	4	4	4
O/A ratio: top	8·5	6·24 **2**	6·54	5·8	6·9	7·16
Front tyre (in.)	2·25 x 19	3·00 x 18	2·75 x 21	3·25 x 18	3·00 x 18	2·75 x 21
Rear tyre (in.)	2·25 x 19	3·00 x 18	4·00 x 19	3·25 x 18	3·00 x 18	4·00 x 19
Front brake dia. (in.)	4	4	5	6	5	6
Rear brake dia. (in.)	5	5	5	6	5	5
Front suspension	teles	teles	teles	teles	teles	teles
Rear suspension	s/a	s/a	s/a	s/a	s/a	s/a
Petrol tank (Imp. gal)	2·25	2·25	2·25	2·75	2·25	
Wheelbase (in.)	49·5	49·5	51	52	49·5	
Ground clear. (in.)	5	5	6·5	5·5	5	8·5
Seat height (in)	29	28·5	30	30	29	
Dry weight (lb)	165	185	206	280	240	

1 1957—6F option, 1962—6F only **2** 1957—6·6

Model	L25T Commando	L25S Cotswold	L15A Flying Cadet	L20 Captain	L20S Sports Captain	M25 Superswift
Year from	1958	1958	1959	1959	1961	1961
Year to	1962	1962	1962	1966	1966	1963
Engine make	AMC	AMC	AMC	AMC	AMC	Villiers
Engine type	250	250	150	200	200	2T
Gearbox type	4	4 close	3	4	4	4
O/A ratio: top	8·6	8·0	6·7	6·63	6·63	6·2
Front tyre (in.)	2·75 × 21	2·75 × 21	3·00 × 18	3·00 × 18	3·00 × 18 **1**	3·25 × 18
Rear tyre (in.)	4·00 × 19	3·50 × 19	3·00 × 18	3·25 × 18	3·25 × 18	3·25 × 18
Front brake dia. (in.)	6 **2**	6	4	5	6	6
Rear brake dia. (in.)	6	6	5	5	5	6
Front suspension	teles	teles	teles	teles	teles	teles
Rear suspension	s/a	s/a	s/a	s/a	s/a	s/a
Petrol tank (Imp. gal)	2·25	2·25	2·25	2·75	2·75	2·75
Wheelbase (in.)	54·5 **3**	54·5	49·5	51·2	51·2	52
Ground clear. (in.)			5	5	5	5·5
Seat height (in.)	34 **4**	34	30	30	30	32
Dry weight (lb)	300 **5**	300	171	262	260	300

1 1964—2·75 × 19 **2** 1962—5 **3** 1962—53·5 **4** 1962—31·5 **5** 1962—265

Model	M15 Cadet	M25T Commando	M25R Cotswold	M25S Sports Superswift	M25RS Cotswold	Model M16
Year from	1962	1962	1962	1962	1964	1965
Year to	1965	1966	1964	1966	1966	1966
Engine make	AMC	Villiers	Villiers	Villiers	Villiers	AMC
Engine type	150	32A	36A + Parkinson	2T **1**	Starmaker	150
Gearbox type	3	4	4 close	4	4	3
O/A ratio: top	6·7	6·88	8·6	6·21	10·75	6·7
Front tyre (in.)	3·00 × 18	2·75 × 21	2·75 × 21	3·25 × 18 **2**	2·75 × 21	3·00 × 18
Rear tyre (in.)	3·00 × 18	4·00 × 19	4·00 × 18	3·25 × 18	4·00 × 18	3·00 × 18
Front brake dia. (in.)	5	5	6	6	6	5
Rear brake dia. (in.)	5	5	5	5	6	5
Front suspension	teles	teles	teles	teles	teles	teles
Rear suspension	s/a	s/a	s/a	s/a	s/a	s/a
Petrol tank (Imp. gal)	2·75	2·25	1·5	2·75	1·5	2·25
Wheelbase (in.)	49·5		52·5	51·2		
Ground clear. (in.)	6·5		9	5		
Seat height (in.)	29·5		29	30		
Dry weight (lb)	200	248		290	251	165

1 1964—4T **2** 1964—2·75 × 19

Mercury

Model	Grey Streak
Year	1956
Engine make	Villiers
Engine type	6F
Gearbox type	2
Front tyre (in.)	2·25 × 19
Rear tyre (in.)	2·25 × 19
Front brake dia. (in.)	4
Rear brake dia. (in.)	4
Front suspension	teles
Rear suspension	s/a
Petrol tank (Imp. gal)	2·0

New Hudson

Year from	**1946**	**1949**
Year to	**1949**	**1958**
Engine make	Villiers	Villiers
Engine type	JDL	2F
Gearbox type	1	1
O/A ratio: top		10·75
Front tyre (in.)		2·25 × 21
Rear tyre (in.)		2·25 × 21
Front brake dia. (in.)		4
Rear brake dia. (in.)		4
Front suspension	girder **1**	girder
Rear suspension	rigid	rigid
Petrol tank (Imp. gal)	1·75	1·62 **2**
Wheelbase (in.)		50
Ground clear (in.)		3·5
Seat height (in.)		32·5
Dry weight (lb.)		123

1 from 1948   **2** 1956—1·25

Norman

Model	**Autocycle**	**Model C**	**Model D**	**Motorcycle**	**B1**	**B2**
Year from	**1946**	**1949**	**1950**	**1946**	**1949**	**1949**
Year to	**1948**	**1957**	**1955**	**1948**	**1952**	**1954**
Engine make	Villiers	Villiers	Villiers	Villiers	Villiers	Villiers
Engine type	JDL	2F	1F **1**	9D	10D	6E **2**
Gearbox type	1	1	2	3	3	3
O/A ratio: top		10·75	8·12		7·0	5·86
Front tyre (in.)	2 × 26	2·25 × 21	2·50 × 19		3·00 × 19	3·00 × 19
Rear tyre (in.)	2 × 26	2·25 × 21	2·50 × 19		3·00 × 19	3·00 × 19
Front brake dia. (in.)		4	4		5	5
Rear brake dia. (in.)		4	4		5	5
Front suspension	girder	girder	girder **3**	girder	teles	teles
Rear suspension	rigid	rigid	rigid	rigid	rigid	rigid
Petrol tank (Imp. gal)		1·75	1·75		2·75	2·75
Wheelbase (in.)		50·5	48		49	49
Ground clear. (in.)		4·5	4·5		5	5
Seat height (in.)		32	27·5		27·5	27·5
Dry weight (lb)	110	130	140		172	206

1 1954—4F   **2** 1954—8E   **3** 1951—teles

Model	**BIS**	**E**	**B2S**	**B2C**	**TS**	**B2SC**
Year from	**1952**	**1952**	**1952**	**1952**	**1955**	**1956**
Year to	**1959**	**1953**	**1958**	**1955**	**1957**	**1960**
Engine make	Villiers	Villiers	Villiers	Villiers	Anzani	Villiers
Engine type	13D **1**	10D	6E **2**	6E **3**	242	9E
Gearbox type	3 **4**	3	3 **5**	3 **5**	4	4
O/A ratio: top	6·5 **6**	6·8	5·86 **7**	6·6	5·5	7·1
Front tyre (in.)	3·00 × 19	2·50 × 19	3·00 × 19	3·00 × 19 **8**	3·00 × 19	2·75 × 21
Rear tyre (in.)	3·00 × 19	2·50 × 19	3·00 × 19	3·50 × 19 **8**	3·00 × 19	4·00 × 19
Front brake dia. (in.)	5	4	5	5	5	6
Rear brake dia. (in.)	5	4	5	5	5	6
Front suspension	teles **9**	teles	teles **10**	teles	leading link	leading link

Villiers Singles & Twins

167

Rear suspension	s/a	rigid	s/a **11**	rigid	s/a	s/a
Petrol tank (Imp. gal)	2·25	1·75	2·25	1·75	2·25	1·75
Wheelbase (in.)	50·2	48	49·8	48	50·2	51
Ground clear. (in.)	6·7	4·5	6·7	7	6·5	7·5
Seat height (in.)	29	27·5	29	27·5	29	29
Dry weight (lb)	230 **6**	155	226	190	275	237

1 1955—30C, 1957—31C also **2** 1954—8E **3** 1954—7E **4** 1957—3 or 4 **5** 1954—3 or 4 **6** 1955 on **7** 1954—5·7, 1956—6·2 **8** 1954—3·00 × 21 and 4·00 × 19 **9** 1956—leading link **10** 1955—leading link **11** 1954—Armstrong units

Model	**B3**	**B2S**	**B4C**	**B4C Trials**	**B4C Scrambles**	**B4**
Year from	**1958**	**1959**	**1960**	**1960**	**1960**	**1960**
Year to	**1960**	**1961**	**1961**	**1961**	**1961**	**1962**
Engine make	Villiers	Villiers	Villiers	Villiers	Villiers	Villiers
Engine type	2T	9E	9E	32A	34A	2T
Gearbox type	4	3 or 4	4	4	4	4
O/A ratio: top	6·2	6·2		7·0	8·4	6·2
Front tyre (in.)	3·00 × 19 **1**	3·00 × 19 **1**	2·75 × 21	2·75 × 21	2·75 × 21	3·00 × 19 **1**
Rear tyre (in.)	3·00 × 19	3·00 × 19	4·00 × 19	4·00 × 19	3·50 × 19	3·00 × 19
Front brake dia. (in.)	6			6	6	6
Rear brake dia. (in.)	6			6	6	6
Front suspension	leading link	leading link	leading link	leading link	leading link	leading link
Rear suspension	s/a	s/a	s/a	s/a	s/a	s/a
Petrol tank (Imp. gal)	3·25 **2**	3·25 **2**	1·75	1·75	1·75	3·75
Wheelbase (in.)	52·5	52				
Ground clear. (in.)	6					
Seat height (in.)	30	31				
Dry weight (lb)	307 **3**	272 **4**		237	237	304 **5**

1 Sports—2·75 × 19 **2** Sports—2·5 **3** Sports—302 **4** Sports—252 **5** Sports—294

OEC

Model	**S1 1**	**C1**	**SS1 2**	**D55**	**D55RS**
Year from	**1949**	**1950**	**1951**	**1953**	**1953**
Year to	**1953**	**1952**	**1953**	**1954**	**1954**
Engine make	Villiers	Villiers	Villiers	Villiers	Villiers
Engine type	10D	10D	10D	10D	10D
Gearbox type	3	3	3	3	3
O/A ratio: top	7·55	9·8	7·55	7·55	7·55
Front tyre (in.)	2·75 × 19	2·75 × 19	2·75 × 19	2·75 × 19	2·75 × 19
Rear tyre (in.)	2·75 × 19	3·25 × 19	2·75 × 19	2·75 × 19	2·75 × 19
Front brake dia. (in.)	4	4	4		
Rear brake dia. (in.)	5	5	5		
Front suspension	teles	teles	teles	teles	teles
Rear suspension	rigid	rigid	s/a	rigid	s/a
Petrol tank (Imp. gal)	2·5	2·5	2·5	2·25	2·25
Wheelbase (in.)	50	50			
Ground clear. (in.)	5·5	7·5			
Seat height (in.)	27	31			
Dry weight (lb)	180	170	190	170	180

1 and D1 from 1950 **2** and SD1 from 1951

Villiers Singles & Twins

Model	S2 1	C2	SS2 2	ST2	ST3
Year from	1949	1950	1951	1953	1953
Year to	1953	1952	1953	1954	1954
Engine make	Villiers	Villiers	Villiers	Villiers	Villiers
Engine type	6E	6E	6E	6E 3	6E 4
Gearbox type	3	3	3	3	3 or 4
O/A ratio: top	5·33	6·9	5·33	5·33	5·29
Front tyre (in.)	3·00 × 19	2·75 × 19	3·00 × 19	3·00 × 19	2·75 × 19 or 21
Rear tyre (in.)	3·00 × 19	3·25 × 19	3·00 × 19	3·00 × 19	3·25 × 19
Front brake dia. (in.)	4	4	4	5	5
Rear brake dia. (in.)	5	5	5	5	5
Front suspension	teles	teles	teles	teles	teles
Rear suspension	rigid	rigid	s/a	s/a	s/a
Petrol tank (Imp. gal)	2·5	2·5	2·5	2·75	2·75 or 1·5
Wheelbase (in.)	50	50		51	
Ground clear. (in.)	5·5	7·5		6	7·5
Seat height (in.)	27	31		29	
Dry weight (lb)	190	180	200	205	210

1 and D2 from 1950 **2** and SD2 from 1951 **3** 1954—8E **4** 1954—7E

Panther

Model	10/3	10/4	10/3A	35	35 Sports	45 Sports
Year from	1955	1955	1956	1956	1957	1958
Year to	1960	1962	1962	1960	1962	1964
Engine make	Villiers	Villiers	Villiers	Villiers	Villiers	Villiers
Engine type	8E	9E	9E	2T	2T	3T
Gearbox type	3	4	3	4	4	4
O/A ratio: top	5·8	6·2	6·2	6·2	6·2	5·9
Front tyre (in.)	3·25 × 18	3·25 × 18	3·25 × 18	3·25 × 18	3·25 × 18	3·25 × 18
Rear tyre (in.)	3·25 × 18	3·25 × 18	3·25 × 18	3·25 × 18	3·25 × 18	3·25 × 18
Front brake dia. (in.)	5 1	5 1	5 1	6	7	7
Rear brake dia. (in.)	5 1	5 1	5 1	6	6	6
Front suspension	Earles	Earles	Earles	Earles 2	Earles 2	Earles 2
Rear suspension	s/a	s/a	s/a	s/a	s/a	s/a
Petrol tank (Imp. gal)	2·5	2·5	2·5	2·5	2·5	2·5
Wheelbase (in.)	52·5	52·5	52·5	52·5	52·5	52·5
Ground clear. (in.)	6	6	6	66	6	
Seat height (in.)	28·5	28·5	28·5	28·5	28·5	28·5
Dry weight (lb)	235	245	235	290		309

1 1957—6 **2** 1960—teles

Model	50 Grand Sports	35
Year from	1959	1962
Year to	1962	1968
Engine make	Villiers	Villiers
Engine type	3T	2T
Gearbox type	4	4
O/A ratio: top	5·9	6·2
Front tyre (in.)	3·25 × 18	3·25 × 18
Rear tyre (in.)	3·25 × 18	3·25 × 18

Front brake dia. (in.)	8	7
Rear brake dia. (in.)	8	6
Front suspension	teles	teles
Rear suspension	s/a	s/a
Petrol tank (Imp. gal)	2·5	2·5
Wheelbase (in.)	52·5	52·5
Ground clear. (in.)	6	6
Seat height (in.)	28·5	28·5
Dry weight (lb)	333	290

Radco

Model	**Ace**
Year	**1954**
Engine make	Villiers
Engine type	4F
Gearbox type	2
O/A ratio: top	8·47
Front tyre (in.)	2·50 × 19
Rear tyre (in.)	2·50 × 19
Front brake dia. (in.)	4
Rear brake dia. (in.)	4·6
Front suspension	leading link
Rear suspension	rigid
Petrol tank (Imp. gal)	1·75
Dry weight (lb)	138

Rainbow

Year	**1950**
Engine make	Villiers
Engine type	1F
Gearbox type	2
Front tyre (in.)	26 × 2
Rear tyre (in.)	2·50 × 20
Front suspension	girder
Rear suspension	rigid
Petrol tank (Imp. gal)	0·75
Wheelbase (in.)	44
Dry weight (lb)	112

Scorpion

Model	**Trials**	**Scrambler**	**Trials Mk 2**	**Moto-Cross Mk 4**	**Racer 250**
Year	**1963**	**1963**	**1964**	**1964/65**	**1965**
Engine make	Villiers	Villiers	Villiers	Villiers	Scorpion
Engine type	9E or 32A **1**	34A	32A	36A or Starmaker **2**	250
Gearbox type	4	4	4	4	5
O/A ratio: top	6·7				
Front tyre (in.)	2·75 × 21	3·00 × 21			
Rear tyre (in.)	4·00 × 18	3·50 × 19			
Front brake dia. (in.)	6	6	6	6	200 mm

Rear brake dia. (in)	6	6	6	6	200 mm	
Front suspension	leading link	leading link	leading link	leading link **3**	teles	
Rear suspension	s/a	s/a	s/a	s/a	s/a	
Petrol tank (Imp. gal)	1 or 2			1 **4**	1·5	
Wheelbase (in.)	51·5	52	52	52·5		
Ground clear. (in.)	9·5	8	9·5	8		
Seat height (in.)	30					
Dry weight (lb)	200	211	215	211		

1 or Starmaker or converted   **2** 1965—Scorpion unit   **3** 1965—teles for Scorpion powered unit   **4** 1965—1·5

Sprite

Model	Scrambles	Trials	Monza trials	Monza scrambler	Sprite 360	405 Moto-Cross
Year from	**1964**	**1964**	**1966**	**1966**	**1968**	**1969**
Year to	**1967**	**1967**	**1967**	**1967**	**1968**	**1974**
Engine make	Alpha/Greeves	Villiers	Villiers	Villiers	Sprite	Sprite
Engine type	250 cc	32A	Starmaker	Starmaker	360	400
Gearbox type	4-Albion	4	4	4	4	4
O/A ratio: top	optional	optional	optional	optional	optional	8·2
Front tyre (in.)						3·00 × 21
Rear tyre (in.)						4·00 × 18
Front brake dia. (in.)						6
Rear brake dia. (in.)						6
Front suspension	teles **1**	teles **1**	leading link	leading link	leading link	leading link
Rear suspension	s/a	s/a	s/a	s/a	s/a	s/a
Petrol tank (Imp. gal)	1·5	1·5 **2**	1·75			1·25 or 2·75
Wheelbase (in.)						53·5
Ground clear. (in.)			10·5			7·75
Seat height (in.)			29			32
Dry weight					210	211

1 1965—leading link   **2** 1965—2

Model	250 Moto-Cross
Year from	**1969**
Year to	**1974**
Engine make	Sprite
Engine type	250
Gearbox type	4
O/A ratio: top	optional
Front tyre (in.)	3·00 × 21
Rear tyre (in.)	4·00 × 18
Front brake dia. (in.)	6
Rear brake dia. (in.)	6
Front suspension	teles
Rear suspension	s/a

Sun

Model	Autocycle	Motorcycle 1	122 de luxe	Challenger de luxe	Challenger comp.	Cyclone
Year from	**1946**	**1949**	**1951**	**1951**	**1952**	**1953**
Year to	**1950**	**1958**	**1954**	**1957**	**1955**	**1957**
Engine make	Villiers	Villiers	Villiers	Villiers	Villiers	Villiers
Engine type	JDL **2**	1F **3**	10D **4**	6E **5**	6E **5**	1H
Gearbox type	1	2	3 **6**	3 **6**	3 **6**	4

Part Three Appendix

O/A ratio: top		8·4		6·0	6·8	6·22
Front tyre (in.)	2·25 × 21	2·50 × 19	3·00 × 19	3·00 × 19	2·75 × 21	3·25 × 19
Rear tyre (in.)	2·25 × 21	2·50 × 19	3·00 × 19	3·00 × 19	3·50 × 19	3·25 × 19
Front brake dia. (in.)		4	5	5	5	6
Rear brake dia. (in.)		4	5	5	5	6
Front suspension	girder	girder	teles	teles	teles	teles
Rear suspension	rigid	rigid	plunger **7**	plunger **7**	plunger **8**	s/a
Petrol tank (Imp. gal)		1·5 **9**	2·25	2·25 **10**	2·25	2·5 **11**
Wheelbase (in.)		48		48	49	51
Ground clear. (in.)		5		7·5	7·5	7·5
Seat height (in.)		26·5		28·5	32	31
Dry weight (lb)		136		214	190	264

1 1955—Hornet **2** 1949—2F **3** 1953—4F **4** 1953—12D **5** 1953—8E **6** 1954—3 or 4 **7** 1954—s/a **8** 1954—rigid
9 1955—2·25 **10** 1956—2·5 **11** 1956—2·75

Model	**Challenger Mk IA**	**Scrambler**	**Wasp**	**Wasp comp.**	**Century**	**Wasp Twin**
Year from	**1954**	**1954**	**1955**	**1955**	**1956**	**1956**
Year to	**1957**	**1955**	**1959**	**1957**	**1957**	**1959**
Engine make	Villiers	Villiers	Villiers	Villiers	Villiers	Villiers
Engine type	30C **1**	8E	9E	9E	8E	2T
Gearbox type	3 or 4	3 or 4	4 **2**	4	3	4
O/A ratio: top	6·5	6·8	6·2	6·2		6·2
Front tyre (in.)			3·00 × 19			3·00 × 19
Rear tyre (in.)			3·00 × 19			3·00 × 19
Front brake dia. (in.)			5	5	5	5
Rear brake dia. (in.)			5	5	5	5
Front suspension	teles	leading link	leading link	leading link **3**	teles	leading link
Rear suspension	s/a	s/a	s/a	s/a	s/a	s/a
Petrol tank (Imp. gal)	2·2—	2·25	2·5	2·5	2·75	2·75
Wheelbase (in.)	48	49·5	52	51		54
Ground clear. (in.)	7·5	9	7	8		7
Seat height (in.)	28·5	29	26		26	
Dry weight (lb)	208	220	244			296

1 1957—31C **2** 1957—3 or 4, 1958—3 only **3** 1957—or Earles

Tandon

Model	**Milemaster Mk I**	**Superglide Mk II**	**Superglide Supreme**	**Kangaroo**	**Kangaroo Supreme**	**Imp**
Year from	**1948**	**1950**	**1951**	**1951**	**1952**	**1953**
Year to	**1952**	**1953**	**1953**	**1953**	**1953**	**1955**
Engine make	Villiers	Villiers	Villiers	Villiers	Villiers	Villiers
Engine type	9D	10D	6E	10D	6E	10D **1**
Gearbox type	3	3	3	3	3	3
O/A ratio: top	7·7	7·55	5·32	9·05		7·55
Front tyre (in.)	2·50 × 19	3·00 × 19	3·00 × 19	2·75 × 21	2·75 × 21	3·00 × 19
Rear tyre (in.)	2·50 × 19	3·00 × 19	3·00 × 19	3·25 × 19	3·25 × 19	3·00 × 19
Front brake dia. (in.)	4	4 **2**	5	5	5	5
Rear brake dia. (in.)	4	4 **3**	5	5	5	5
Front suspension	teles	teles	teles	teles	teles	teles
Rear suspension	rigid	s/a	s/a	s/a	s/a	rigid
Petrol tank (Imp. gal)	2·5	2·5	2·5	1·0	1·0	1·5 **4**
Wheelbase (in.)	48	50	50	48	48	48
Ground clear. (in.)	4·5	6·75	6·75	7·5	7·5	6
Seat height (in.)	27·5–29	27	27	31	31	27
Dry weight (lb)	154	160	187	168		168

1 1954—12D, 1955—30C **2** 1953—5 **3** 1951—5 **4** 1955—1·75

Tandon

Model	Imp Supreme	Scrambler	Monarch	Twin Supreme	Viscount
Year from	1953	1954	1954	1954	1955
Year to	1959	1955	1959	1955	1955
Engine make	Villiers	Villiers	Villiers	Anzani	Anzani
Engine type	6E **1**	8E **2**	1H	250	325
Gearbox type	3 **3**	4	4	4	4
O/A ratio: top	5·88	optional	6·07	6·33	5·1
Front tyre (in.)	3·00 × 19	3·00 × 19	3·00 × 19	3·00 × 19	3·00 × 19
Rear tyre (in.)	3·00 × 19	3·25 × 19	3·00 × 19	3·00 × 19	3·00 × 19
Front brake dia. (in.)	5	5	5 **4**	5 **4**	6
Rear brake dia. (in.)	5	5	5 **4**	5 **4**	6
Front suspension	teles **5**	Earles	teles **6**	teles **6**	leading link
rear suspension	s/a	s/a	s/a	s/a	s/a
Petrol tank (Imp. gal)	1·75 **7**	2·0	2·25 **8**	2·25 **8**	2·5
Wheelbase (in.)	49				49·5
Ground clear. (in.)	7	8·5			6·5
Seat height (in.)	27				27·5
Dry weight (lb)	176		209	209	262

1 1954—8E **2** 1955—7E **3** 1954—or 4 **4** 1955—6 **5** Special 1955—leading link **6** 1955—leading link **7** de luxe—2·25 **8** 1955—2·5

Engine and frame numbers

Aberdale

Year	Model	Engine	Frame
1949	Autocycle	A608/21493	AAU 763

Ambassador

Year	Model	Engine	Frame
1946	197 Series I	662-155	1116103
1947	197 Series I	662-197	1027141
1948	197 Series I	662-479	1018423
	197 Series II	662-495	2038449
	197 Series III	809-154	3058595
1949	197 Series III	809-1450	3291375
1950	197 Series III	944-15484	44503266
	197 Series IV	944-6628	3892050
	197 Series V	949-6435	3892093
1951	197 Series III	949-27261	311504711
	197 Series IV	944-28024	410504537
	197 Series V	949-28036	E1250-4884
	197 Supreme	139A-35116D	S3515396
1952	197 Courier	231A-45924D	C9516786 (last)
	197 Popular	227A-43234D	P6516181
	197 Embassy	231A-43222D	E7516359
	197 Supreme	139A-37028D	S7516273
1953	197 Popular	363A-12305	P95310603
	197 Embassy	363A-12265	E95310592
	197 Supreme	139A-69615	S8528945
	197 SS Supreme	654A-5428	S4539911
	197 Sidecar	231A-54483	SX1527545
1954	197 SS Embassy	363A-15731	SS95310736
	197 Supreme	801A-15718	S95310564 (last)
	197 Embassy	801A-20754	E125311254

Year	Model	Engine	Frame
	197 Popular	810A-26318	P35411551
	197 Sidecar	654A-5453	X65310277
	224 Supreme	935A-396	S35411589
1955	197 Popular	363A-24775	PM54-12192
	197 Embassy	045B-37620	EB54-12286
	197 Embassy	801A-33051	EB54-12279
	197 SS Embassy	045B-55695	SS954-12432
	197 Sidecar	363A-33990	X854-12118
	197 Envoy	045B-39785	1154-12840
	224 Supreme	935A-2084	S85412118
1956	147 Popular	030B-12185	P2/56
	197 Embassy	064B-51283	E45513708
	197 Sidecar	045B-55695	X55513914 (last)
	197 Envoy	045B-64040	K105514187
	197 Envoy	013B-719	K105514192
	197 Popular	045B-61624	P555-13889 (last)
	224 Supreme	935A-11000	S105514183
	197 Embassy	045B-53787	E45513713 (last)
1957	147 Popular	030B-14537	P75614738 (last)
	148 Popular	014-172	P10561-4811
	148 Popular	015-167	P10561-4812
	197 Envoy	018-247	K105614816
	197 Envoy	019-2884	K105614815
	224 Supreme	935A-9880	S75614765 (last)
	249 Supreme Twin	021-184	S75614768
1958	148 Popular	525B-438	P75715427
	197 Envoy	P018-248	K85715535
	197 Envoy	313B-171	K85715536
	246 Supreme	543B-163	S25816115
	249 Supreme Twin	950A-1782	S95715572
1959	148 Popular	525B-941	P75816561 (last)
	173 Statesman	512B-414	T45816260 (last)
	173 Popular	512B-435	P115816747
	197 Envoy	580B-1840	K15917003 (last)

Year	Model	Engine	Frame	Year	Model	Engine	Frame
	197 Envoy	770B-4853	K15917015 (last)	1950	Mk II	24146	15775
	197 3 Star Special	580B-1961	45917374	1951	Mk II	29562	21258
	246 Supreme	543B-155	S55816388 (last)	1952	Mk IV	1008	25008
	249 Supreme Twin	950A-3754	S75816494 (last)	1953	Mk IV	2101	26101
	249 Super S	808B-5347	S95816571	1954	Mk IV	2604	26604
1960	197 3 Star Special	580B-1225	X95917784	1954	Mk IV	3050	27050 (last)
	197 3 Star Special	770B-6238	X95917871	**Cotton**			
	249 Super S	808B-6350	S95917912	1956	197 Vulcan	57679E	5613
1961	173 Popular	512B-526	(last)		242 Cotanza	2135	A5610
	197 3 Star Special	580B-3459	X66019458		322 Cotanza	2038	A5612
	197 3 Star Special	770B-8397	X1061190	1957	197 Vulcan	146E-78487	5767
	249 Electra 75	208D-816	E96124		242 Cotanza	2808	5753
	249 Super S	808B-9668	S76019571		249 Villiers Twin	950A-178	5757
	249 Super Sports	269D-10029	SS1161375		322 Cotanza	3149	A5649
1962	197 3 Star Special	580B-5183	X661715	1958	197 Vulcan	228B-446	58115
	197 3 Star Special	770B-9171	X661713		249 Villiers Twin	950A-1563	58112
	197 Popular	507D-9497	PWTNW-52		322 Cotanza	4050	58202
	249 Super S	214D-11680	S661643	1959	197 Vulcan	544B-535	58220
	249 Super Sports	269D-10434	SS661670		197 Trials	478B-3626	58162
1964	249 Super S		2M-21		246 Scrambler	504B-4920	T18
1965	197 3 Star Special		9M-39 (last)		249 Herald	950A-2974	58218
	249 Super S		2M-39 (last)		322 Cotanza	4101	59247
BAC					324 Messenger	847B-180	59227
1951	98 Lilliput	716-11960	X102	1960	197 Vulcan	544B-2309	60470
	125 Lilliput	F2U-H	T100		197 Trials	863B-789	T26
		80391-2			242 Cotanza	C1-4410	60529
Bond					246 Scrambler	016D-541	T34
1950	98 Minibyke	716-8969	A101		249 Herald	039D-6126	60459
	98 Minibyke	161A-	50V		322 Cotanza	C2-4416	60527
1951	98 Minibyke	161A-	51V		324 Messenger	054D-772	60458
	125 Minibyke	F2U/H	51J	1961	197 Trials	150D-7996	T134
1952	98 Minibyke	161A-	52V		197 Vulcan Sports	313B-8373	61722
	125 Minibyke	F2U/H	52J		197 Vulcan	544B-3754	60680
1953	98 Minibyke	161A-19695	53V (last)		246 Scrambler	070D-1452	T140
	125 Minibyke	F2U/H80422/2	53J (last)		246 Trials	149D-1689	T146
Bown					249 Herald	039D-6546	60675
1950	98 Auto Roadster	801-7393	BAU1876		249 Double Glos.	039D-6562	60676
1951	98 Standard		BMC1001		249 Continental	699B-10144	61C1
1952	98 Auto Roadster	801-6590	BAU1901		324 Messenger	054D-744	60697
	98 Standard	716-18723	BLM1	1962	197 Vulcan Sports	313B-9264	61839
Corgi					197 Trials	150D-9247	T326
1948	Mk I	M5385	2372		246 Corsair	400D-176	61810
	Mk II	16005	6303		246 Trials	149D-2298	T285
					246 Scrambler	248D-2193	T283
					246 Cougar	284D-1231	CT3
					246 Corsair	400D-170	61A2

Villiers Singles & Twins

175

Part Three Appendix

Year	Model	Engine	Frame
	249 Herald	004D-9121	61813
	249 Double Glos.	004D-9071	61812
	249 Continental	393D-11950	61C125
1963	197 Vulcan	544B-6640	63878
	197 Vulcan Sports	004E-9046	63879
	246 Cougar	086E-151	CT89
	246 Trials	607D-2890	TC7
	246 Trials Special	149D-2347	TC2
	246 Scrambler	284D-2213	T304 (last)
	247 Cobra	757D-170	C1
	249 Herald	725D-13685	62876
	249 Double Glos.	004D-9885	62866
	249 Continental	606D-13266	62C184
	249 Continental Sports	229D-11576	62C151
	324 Messenger	054D-1506	62C202
1964	246 Corsair	863E-183	638106 (last)
	247 Telstar	225H/490E/708	RR27 (last)
	247 Telstar	2225H/834E/634	RR28
	249 Continental		63C234 (last)
	249 Continental Sports		63C237
	249 Herald	229D-10741	63893 (last)
	324 Messenger	353E-1698	65C283 (last)
1965	197 Vulcan	040D-10789	65913
	246 Cougar	086E-1640	CT334
	246 Trials	607D-3784	TC119
	246 Trials Special	607D-3783	TC123
	247 Cobra	826E-803	C190
	247 Telstar	834E-814	RR42
	247 Trials (Star)	871-816	TC122
	247 Conquest	972E-815	CQ65/1
	249 Continental Sports	353E-1602	65C263
1966	197 Vulcan	251D-10318	65918
	246 Trials	607D-3556	TC178
	246 Trials Special	607D-3556	TC129
	247 Telstar	834E-820	RR69
	247 Trials (Star)	131F-1024	TC190
	247 Conquest	972E-1094	CQ65/3
	247 Cobra	826E-852	CA24
	249 Continental Sports	353E-1698	65C283
1967	246 Cougar	086E-1557	CA82 (last)
	246 Trials 37A	161F-685	T627
	247 Trials (Star)	131F-1187	T697Y
1968	246 Trials 37A	161F-685	T627

Year	Model	Engine	Frame
	247 Trials (Star)	131F-1024	TC190
	247 Conquest	972E-1094	CQ65/3
	247 Cobra	826E-852	CA24
	249 Continental Sports	687E-1876	67C301

Cyc-Auto

Year	Model	Engine	Frame
1946	98 Autocycle	4608	
1947	Autocycle	5198	
1948	Autocycle	6584	
1949	Autocycle	7364	
1950	98 Superior	7894	
1951	Superior	8077	B155
1954	Superior	8132	B310
1955	Superior	8175	B2035
1956	Superior		B3000

DMW

Year	Model	Engine	Frame
1951	125 std.	941-14938D	R5018
	125 dl	945-28478D	S5190
	125 comp.	036A-180980	R5114
1952	125 std.	218A-280130	R52332
	125 dl	218A-29425	S52401
	125 comp.	218A-280150	S52381
1953	125 std.	392A-1843	R531041
	125 dl	218A-29423	S53852
1954	225 Cortina		2P108
1955	147 Leda	133B-268	1P655L
	197 M-X	659A-2677	SP195
	197 200P	053B-39543	1P512
	225 Cortina	993A-4322	2P217
1956	147 Leda	133B-584	1P11436
	197 Mk 7 Trials	415B-706	7P101
	197 M-X Mk 6	414B-800	6P101
	197 200P, Mk 1	053B-45162	1P1152
	197 200P, Mk 9	326B-402	9P103
	225 Cortina	993A-5735	2P761
1957	148 150P, Mk 9	083-183	9P297E
	173 175P, Mk 9	634B-155	9P355E
		634B-155	9P355E
	225 Cortina	993A-5151	2P945E (last)
	249 Dolomite II	563-722	D2P1E
1958	197 200P, Mk 1	044B-80060	1P1313 (last)
	197 M-X, Mk 6	415B-1526	6P130E (last)
	197 Trials Mk 7	414B-2990	7P180E (last)
	249 Dolomite II		D2P81
	249 Mk 10		10P1EB

176

Year	Model	Engine	Frame	Year	Model	Engine	Frame
1962	246 Mk 14 Scrambler	284D-2440	14-1		98 F4F	606B-9125	F4F/10188
	246 Mk 15 Trials	161D-1422	15-1		147 U9	356B-32000	PU400
					147 U9R	356B-32080	PU500
Excelsior					246 TT4	R10410	7PT/495
1954	98 F4	652A-2165	F4/100		328 S8	R10418	8ST510
1956	98 F4	652A-13879	F4S/9313	1960	98 F4F	606B-12950	10799
	98 F4S	652A-18059	F6S/100		98 F10	606B-12330	F/10800
	98 G2	5875	AX6849		98 C10	606B-12300	CC1080
	98 S1	S/7956	SB6658		147 U9	356B-31950	U9/1340
	148 C3	C1906	5PC/810		147 U9R	356B-31960	U9/1341
	148 C4	C2105	6PC/1133		148 U10	051D-1000	U/100
	197 R4	049B-44571	5AR/839		197 R10	059D-6000	R/100
	197 R5	049B-42731	5PR/743		246 TT4	R10810	9P/8000
	197 R6	049B-41655	5PR/661		246 TT6	R10810	9/800
	197 A9	313B-166	5PT/566		328 S8	R10550	8S/500
	246 TT2	7180	5AT/1062		328 S9	R10800	9/160
	246 STT2	7240	5AT/1082	1961	98 EC11	606B-15285	SC9/1577
	246 SE-STT2	7344	5AT/1072		197 ER11	059D-7546	RT10/719
	246 TT3	7167	5PT/393		246 ETT7	R13431	9PT/1085
	246 STT4	7517	5AT/1083		246 TT7	R13303	9PT/1084
1957	98 F4	350	9700	1962	98 C11	606B-15948	SC9/1816
	98 F4S	299	949		98 EC11	606B-15944	CS9/1880
	98 F6S	300	950		98 F12	606B-21858	F4S/12868
	98 SB1	470	100		98 C12	606B-15646	SC9/1877
	98 G2	5960	6940		148 U11	051B-2306	U10/1019
	98 S1	7750	6745		148 U12	C131	U10/1101
	148 C4	2340	1270		197 R11	059D-7682	RT10/T44
	148 C3	2250	1120 (last)		197 ER11	059D-8001	RT10/747
	148 C4	C2520	6PC/1500 (last)		246 ETT8	R13505	9PT/1264
	197 R6	049B-72727	6PR/1018		328 S10	R12424	T9/200
	246 TT3	7650	870	1963	98 F12	730D-170	F4S/12923
	246 STT4	7500	900		98 C14	730D-300	SC9/1980
	246 TT4	R5000	7PT/100		148 U14	051D-2399	U63/100
1958	98 F4	605B-21887	F4S/10078		148 U12	C250	U11/1150
	98 F6S	605B-4989	F6S-2000		246 ETT8	R13540	
	98 SB1	605B-5200	SK1500		328 ETT9	R13530	T9/300
	98 CA8	605B-6520	SAC/100	1964	98 C14	606B-17085	SC9/2152
	197 R6	049B-81050	6PR/1070	1965	148 U14	051D-4601	U63/369X1
	197 A9	270B-999	7PT/290				
	246 STT5	8906	274	**Francis-Barnett**			
	246 TT4	R9000	7PT/490	1946	98 50	XXA	LH
	246 STT6	R9100	7PT/300		122 51	597	LK
	328 S8	R9160B	7ST300	1947	98 50	XXA 434	MH
1959	98 F6S	606B-8356	F6S/2402		122 51	597	LK
	98 CA9	606B-8819	SAC/429	1948	98 50	434	NH
	98 SB1	606B-6799	SK1984		122 51	597	NK
				1949	98 50	434	OH

Villiers Singles & Twins

177

Part Three Appendix

Year	Model	Engine	Frame	Year	Model	Engine	Frame
	98 56	801	ON		225 75	842A	YB
	122 51	597-40195	OK-45072 (last)		249 80	25T	YC
	122 52	838	OL	1958	147 78	295B	Z
	197 54	824	OM		171 79	17T-1357	ZD
1950	98 50	434-23483	OH-13800 (last)		197 81	662B	ZN
	98 56	801	PN		249 80	25T13613	ZC
	122 53	823	OL		249 82	25/S-16144	ZCS
	122 52/53	935	PPL	1959	147 78	295B	A
	197 54/55	946	PPM		171 79	17T-1867	AD
1951	98 56	801	RN		197 81	662B	AN
	98 56	189A	RN		249 82	25/S-16168	ACS
	122 52/53	935	RL		249 83	25/C-16303	ACT
	122 52/53	206A	RL		249 80	25T/80-15983	AC
	197 54/55	946	RM		249 84	25T-17153	AE
	197 54/55	207A	RM	1960	149 86	15T-1580	B
1952	98 56	189A	SN		171 79	17T-3906	BD
	122 57	208A	SLS		197 81	662B-6834	AN88117 (last)
	122 59	166A	SLC		199 87	20T	
	197 54/55	207A	SM		249 80	M25T/80-	
	197 58	209A	SMS			18106	BC
	197 60	158A	SMC		249 84	M25T/84-	
1953	122 59	166A-33526	SLC60949 (last)			17995	BE
	122 52/53	206A	TL		249 85	25/C	
	122 57	208A	TLS	1961	149 86	15T-5364	C
	122 61	420A	UTL		171 79	17T-4416	BBD (last)
	122 63	403A	TTLS		199 87	20T-4488	CF
	197 54	207A	TM		249 80	M25T-20243	CC
	197 58	209A	TMS		249 84	M25T-20070	CE
	197 60	158A	SMC	1962	149 86	15T-8433M	D
	197 55	387A-2766	TM65095 (last)		149 88		DGR
	197 62	375A	UTM		199 87	20T-7071	DF
	197 64	374A	UTM		249 80	M25T-20573	DC
1954	122 66		U1001		249 84	M25T/84-	
	197 67	387A	UM			20133	DE
	225 68	842A	UB		249 85	M25C-18846	DCT
1955	147 69				249 89	429D	DH
	225 71	842A	VB	1963	149 90		E90-101
1956	147 73		W1001		199 87		EF
	197 74		WM		246 92		ECT
	197 76		WTM		246 93		ECS
	197 77		WSM		249 80	25T-20899	DC (last)
	225 75	842A	WB		249 89		EH
1957	147 73	295B	W		249 91		E91-101
	147 78	295B	Y	1964	149 95	V15T-10818	F95
	197 74		YM		199 87	V20T-8263	FF
	197 76		YTM		246 92	326E	F92
	197 77		YSM		246 93	352E	F93

Year	Model	Engine	Frame	Year	Model	Engine	Frame
	246 93		GM25R-391 (last)		197 20R4	155B	7160/R4
	247 94	2225H-826E-808	H94		242 25R	2548	7351/25R
	249 89	687E	FP		249 25D	950A	7202
	249 91	687E	F91		322 32D	2902	7401
1965	149 88	V15T-10097	HM88	1958	197 20R4	155B	7169/R4
	149 90	V15T-11304	H90		197 20S		7528/S
	149 95	V15T-11328	H95		197 20T		7554/T
	199 87	V20T-9207	HF		197 20SA	3171	8201/SA
	246 92	326E	H92		197 20TA	3329	8001/TA
	247 94	2225H-826E-900	H94		197 20SAS	626B	8551
	249 89	688E	HH		242 25R	2139	7357/25R (last)
	249 91	687E	H91		249 25D	1041	8451/D
1966	149 96	VI5T-11506	HM96		249 25TA		8651
	149 95		H95-1490 (last)		249 25SA	950A	8751
	149 88		HM88-2116 (last)		32 32D	2850	7418/32D (last)
	149 90		H90-2116 (last)	1959	197 20D	313B	8398/20D
Greeves					197 20SA	626B	8242/SA
1954	197 20R	361A	404R		197 20TA	723B	59/1001
	197 20D	491A	402R		197 20TAS	723B	59/1001
	197 20S	491A	401S		197 20SAS	626B	59/1901
	197 20T	491A	405T		246 24TAS	340P	59/1501
	242 25D	C/1002	406/25D		246 24SAS	888B	59/1701
1955	197 20R	899A	4710R		246 24DB	864B	59/2401
	197 20R3		5151		249 25D	950A	8505/25D
	197 20R4		5201		249 25TA		8731/2574
	197 20D	900A	5051		249 25SA	950A	59/2201
	197 20S	900A	5251		249 25DB	950A	59/2701
	197 20T	900A	5001	1960	197 20TC	625B	60/1000
	242 25D	1041	5301		197 20TCS	723B	60/1000
	242 25R		5351		197 20SCS	626B	60/1805
	322 32D		5401		246 24TCS	863B	60/1400
1956	197 20R4	900A	5222		246 24DB	864B	60/2500
	197 20D	164B	5076		246 24SAS	888B	60/2005 (last)
	197 20R3	074B	6151		246 24SCS	070D	60/2006
	197 20D	324B	6051		249 25DB	950A	60/2800
	197 20S		6251		324 32DB	893B	60/2825
	197 20T	331B	6001	1961	197 20DB	625B	61/2610
	242 25D	1146	5317		197 20TD	625B	61/1000
	242 25R	2625	6351		197 20SCS	626B	61/2500
	322 32D		6401		197 20DC	626B	61/2804
1957	197 20R3	047B	6152/R3		246 24SCS	070D	61/1900
	197 20D	270B	7051/20D		246 24DB	864B	61/2600
	197 20S		7251/S		246 24TDS	723B	61/1300
	197 20T		7001/T		246 24MCS	070D	61/3650
					246 24DC	864B	61/2700
					249 25DC	221D	61/2900
					324 32DC	222D	61/3500

Villiers Singles & Twins

179

Year	Model	Engine	Frame	Year	Model	Engine	Frame
1962	197 20DC	251D	20DC101		246 24RCS	GPA2-140	24RCS102
	197 20TD	625B	20TD101		246 24MX1	GPA1-385	24MX435 (last)
	197 20TE	625B	20TE101		246 24MX2	GPA5-496	24MX501
	197 20SC	625B-SG	20SC101		249 25DC	353E	25DC330
	246 24DC	864B	24DC101		249 25DC Mk II	353E	25DC350B
	246 24SC	284D	24SC101	1966	197 20DC	251D	20DC431
	246 24TD	863B	24TD101		246 24RCS	GPA2-238	24RCS164 (last)
	246 24MCS	305D	24MCS101		246 24TGS	161F	24TGS102
	246 24TE	863B	24TE101		246 24MX3	GPA5	24MX3-1000
	246 24TES		24TES101		246 24RDS	GPA7-101	24RDS-101
	246 24MDS	284D	24MDS 102		246 24RDS	GPA7-159	24RDS-159 (last)
	249 25DC		25DC101		249 25DC Mk II	086F	25DC420B
	249 25DCX	718D	25DCX102	1967	246 24THS	161F	24THSA-103
	324 32DC	222D	32DC101		246 24MX5	GPA5-1537	24MX5A-101
	324 32DCX	222D	32DCX101		246 24RES	GPA7-163	24RES-101
1963	197 20SC	500D	20SC106		364 36MX4		36MX4-101
	246 24MD		24MD101	1968	246 24TJS	161F	24TJS-101
	246 24TE	453D	24TE322		246 24TJ	161F	24TJ-101
	246 24TES	453D	24TES159		246 24MX4		24MX4-101
	246 24RAS	296E	24RAS102		344 35RFS		35RFS-101
	247 24ME	757D	24ME102	1969	246 24TJ	161F	24TJ-214 (last)
	249 25DD	718D	25DD101		246 24TJS	161F	24TJS-297 (last)
	324 32DD	222D	32DD101		246 24MX4	GPC1/2/194	24MX4-186 (last)
1964	197 20TE	251D	20TE118		364 36MX4	GPB2/3/490	36MX4C-690 (last)
	197 20DC	251D	20DC301	1973	246 60K	GPF/14/113	60K409
	246 24MD	086E	24MD174		380 63F	GPG2/225	63F260
	246 24MDS	086E	24MDS1129				
	246 24TE	453D	24TE433	**James**			
	246 24TES	453D	24TES331	1946	98 Autocycle		J9726
	246 24RAS	296E	24RAS171		122 ML		ML12841
	246 24RAS	085E	24RAS196 (last)	1947	122 ML		ML19881
	246 24MX1	GPA1-101	24MX101	1948	98 Comet		F00100
	246 24RBS	GPA2-101	24RBS101		122 ML		ML26921
	247 24ME	757D	24ME204 (last)	1949	98 Comet		06255
	249 25DC	718D	25DC287		98 Autocycle		J15180
	249 25DD	718D	25DD194 Mk 1		122 Standard		D00100
	249 25DD	353E	25DD501 Mk 2		197 6E		D00100
	249 25DCX	353E	25DCX216	1950	98 Autocycle		S17000
	324 32DD	222D	32DD113 (last)		98 Comet		06256
	324 32DCX	222D	32DCX111 (last)		122 Cadet		C00100
1965	197 20DC	251D	20DC398		197 J8		M02750
	246 24TES	453D	24TES585		197 J8 comp		suffix D or RS
	246 24MDS	085E	24MDS1402	1951	98 Autocycle		S19258
	246 24TE	453D	24TE510		98 Comet		10366
	246 24TFS	453D	24TFS102		122 Cadet		C01062
	246 24MX1	GPA1-297	24MX281		197 J7/J8/JS		M05893
	246 24RBS	GPA2-137	24RBS129	1952	98 Autocycle		S20544

Villiers Singles & Twins

Year	Model	Engine	Frame	Year	Model	Engine	Frame
	98 Comet		J15703		249 L25	25T-17904	CL25-2604
	122 Cadet		M01978		249 L25S	25S-76666	CL25S-109
	197 J7/J8/JS		M08879		249 L25T	25C-16430	CL25T-302
1953	98 J1		J1/21317	1961	149 L15A	15T-6059	DL15A-10208
	98 J10/J3/J4	797	J20067		199 L20	20T-16035	DL20-1607
	122 J5/J6	569A	J5/000101		199 L20S	20TS-6498	DL20S-101
	197 J7/J8	365A	A5000101		249 L25	25T-20369	DL25-3025
	197 J7C	381A		1962	98 L1	607B	EL1-3439
	197 J9	319A	J9/000101		149 L15A	15T-8359M	EL15A-11583
1954	98 J11	521A	J11-000101		199 L20	20T-5687	EL20-2401
	122 J5		J5/005001		199 L20S	20TS-7024	EL20S-502
	197 K7, K7C		K7-000101		249 L25	25T-20676	EL25-3308
	197 J9		J9-005001		249 L25S	25S-17246	EL25S-724
	225 K12		K12/000101		249 L25T	25C-16386	EL25T-339
1955	98 J11	521A	55J11/101		249 M25	428D	EM25-104
	122 J5	618A	55J5/101	1963	98 L1	607B	FL1
	147 J15		55J15/101		149 M15		FM15-101
	197 K7	365A	55K7-101		199 L20		FL20-2702
	197 K7C	380A	55K7/101		199 L20S		FL20S-1056
	197 J9		55J9-101		246 M25T		FM25T-363
	225 K12	799A	55K12-101		246 M25R		FM25R-380
1956	98 L1		L1/101		249 M25	428D	FM25-693
	147 L15		56L15/101		249 M25S		FM25S-101
	197 K7		56K7-101	1964	98L1	607B	GL1-3664
	197 K7C		56K7C-101		98 L1	607B	HL1-3712 (last)
	197 K7T		56K7T-101		149 M15	V15T-10774	GM15-588
	225 K12		56K12-101		199 L20	V20T-1232	GL20-2851
1957	98 L1		57L1-101		199 L20S	V20TS-8527	GL20S-1356
	147 L15		57L15-101		246 M25T	326E	GM25T-383
	197 K7	065B	57K7-101		246 M25R	352E	GM25R-385
	197 K7T	298B	57K7T-101		247 M25RS	2225H-826E-721	H25R-401
	225 K12		57K12-101		249 M25S	687E	GM25S-248
	249 L25		57L25-101	1965	149 M15	V15T-11321	HM15-1238
1958	98 L1	310B	AL1-1111		199 L20	V20T-9206	HL20-2982
	147 L15	295LB	AL15-3082		199 L20S	V20TS-9136	HL20S-1557
	197 K7	652B	AK7-101		246 M25T	326E	HM25T-394
	249 L25	25T-11646	AL25-1152		247 M25RS	2225H-826E-809	H25R-420
1959	98 L1	607B	BL1-1473		249 M25S	687E	HM25S-375
	147 L15	295B	BL15-5101	1966	149 M16	V16T	
	171 L17	L17-2225	BL17-801	**Mercury**			
	197 K7	652B	BK7-1928	1957	98 Grey Streak	B1001	M101
	197 K7T	652B	AK7-107	1958	98 Grey Streak	5286	M462
	249 L25	25T-14883	BL25-2073	**New Hudson**			
	249 L25T	25C-16431	BL25T-201	1949	98 Autocycle		ZE101
	249 L25S	25C-16668	BL25S-101	1950	98 Autocycle		ZE1001
1960	98 L1	607B	CL1-1911	1951	98 Autocycle	S13585	ZE4001
	98 L1	602B	CL1-1961				
	149 L15A	15T-1001	CL15A-6958				

181

Part Three Appendix

Year	Model	Engine	Frame
1952	98 Autocycle	171A	ZE6404
1953	98 Autocycle	171A	ZE7600
1954	98 Autocycle	171A	ZE8640
1955	98 Autocycle	171A	ZE10231
1956	98 Autocycle	176B	N1001
1957	98 Autocycle		N4001

Norman

Year	Model	Engine	Frame
1946–49	98 Autocycle	XXA	
1946–48	122 M/C	AAA	
1949	122 B1	843	B1-101
	197 B2	844	B2-103
1950	98 C	801	C101
	122 B1	843	B1-192
	197 B2	844	B2-173
1951	98 C	801	C642
	98 D	983	D150
	122 B1	938	B1-659
	197 B2	947	B2-656
1952	98 C	801	C1363
	98 D	983	D968
	122 B1	938	B1-1003
	197 B2	229A	B2-1717
1953	98 C	801	C1472
	98 D	983	D1269
	122 B1S	221A	B1S-1320
	122 B1	221A	B1-1335
	122 E	221A	E1312
	197 B2	229A	B2-2636
	197 B2S	229A	BS2-2489
	197 B2C	339A	B2C
1954	98 C	801	C1738
	98 D	534A	D1435
	122 B1S	618A	B1S-1673
	122 B1S	221A	B1S-1662
	197 B2	710A	B2-3520
	197 B2C	381A	B2C-3486
1955	98 C	801	C1982
	98 D	534A	D1543
	147 B1S	127B	B1S-1858
	197 B2S	710A	B2S-4483
	197 B2C	381A	B2C-4512
	242 TS	1201	TS4518
1956	98 D	534A	D1780
	98 C	178B	C2434
	147 B1S	127B	B1S-2055
	197 B2S	066B	B2S-5551

Year	Model	Engine	Frame
	197 B2Sdl	375B	B2S-5522
	197 B2C	043B	B2C-5446 (last)
	197 B2C/S	369B	B2C/S-5523
	242 TS	2034	TS5510
1957	98 C	178B	C2591 (last)
	147 B1S		B1S-2352
	197 B2S		B2S-6000
	197 B2C/S	369B	B2S-6007C
	242 TS		TS-6004
1958	242 TS	2747	TS6658 (last)
	249 B3	734B	B3-6686
1959	147 B1S	127B	B1S-2593
	148 B1S/DL	538B	B1S-2591
	197 B2S	732B	B2S-6908
	197 B2C/S	733B	B2S-6919C
	197 B2S/DL	732B	B2S-6916
	249 B3	734B	B3-6911
1960	197 B2C/S	733B	
	197 B2S	732B	
	249 B3	734B	
1961	197 B2S	732B	B2S-8698
	246 B4C	226	8694C
	249 B4	734B	8691
	249 B4S	734B	8689T

OEC

Year	Model	Engine	Frame
1949	122 S1	763	4921
	197 S2	812	491
1950	122 S1,C1	763	50229
	122 D1	932	50326
	197 S2	812	50181
	197 C2,D2	945	50309
1951	122 S1, SS1, D1	932	51500
	122 C1, SD1	932	51500
	197 S2, SS2,	945	51500
	197 D2, SD2, C2	945	51500
1952	122 S1, SS1, D1, SD1, C1	216A	52000
	197 S2, SS2, D2, SD2, C2		52000
1953	122 S1, D1,	216A	531192
	122 SS1, SD1	216A	RS52A-1162
	197 S2, D2, ST2	228A	531182
	197 SS2, SD2	228A	RS52A-1150
	197 ST3	228A	531212
1954	122 D55	389A	541278
	122 D55RS	389A	54RS1280
	197 ST2	361A	R55486

182

Villiers Singles & Twins

Year	Model	Engine	Frame
	197 ST3	380A	5498RS
Panther			
1956	197 10/3	60172E	V006
	197 10/4	313B	V002
1957	197 10/3	72280E	V447
	197 10/4	313B	V569
	197 10/3A	544B	VA001
	249 35	950A	VT001
1958	197 10/3	75182	V833E
	197 10/3A	544B	VA037D
	197 10/4	313B	V539E
	249 35	950A	VT261C
1959	197 10/3	82239	V966F
	197 10/3A	544B	VA073F
	197 10/4	313B	V988D
	249 35	950A	VT266C
	249 35S	950A	VTS281A
	324 45	847B	VTT101A
	324 50	847B	GS101A
1960	197 10/3	355B	V1108G
	197 10/3A	544B	VA104G
	197 10/4	313B	V1110A
	324 45	847B	VTT299B
	324 50	893B	GS200B
1961	197 10/3A	544B	VA158H
	197 10/4	313B	V12201
	249 35	004D	VT419E
	249 35S	004D	VTS554E
	324 45	893B	VTT375B
	324 50	053D	GS322C
1962	197 10/3A	544B	VA215H
	197 10/4	313B	V1268H
	249 35S	004D	VTS597E
	324 50	053D	GS354C
1964	249 35		VTS705
	324 45		VTT712
1966	249 35E	810B	WS464A
1967	249 35	208D	WS530A
1968	249 35	208D	WS605A
Sun			
1946–48	98 Autocycle	451	MC-
1949	98 Autocycle	801	MC-
	98 M/C	716	OMC-H-

Year	Model	Engine	Frame
1951	98 M/C	716	PMC-H
	122 10D	941	PMC-S
	197 6E	419A	PMC-
1952	98 1F	716A	RMC-H-
	122 10D	262A	RMC-S-
	197 6E	489A	RMC-
1953	98 4F	716A	SMC-H-
	122 12D	392A	SMC-S-
	197 6E	339A	SMC-
1954	98 4F	739A	TMC-1-
	122 12D	392A	TMC2S-
	197 6E	046B	TMC1S-
	225 1H	411A	225TMC-
1955	98 4F	739A	VMC-H-
	147 30C	586A	VMC.SA-
	197 8E	046B	VMC1(SA)-
	197 8E/4	055B	VMC
	225 1H	411A	225VMC
1956	98 4F	739A	WMC-H-
	147 30C	586A	WMC.SA
	197 8E	046B	WCMC.SA
	225 1H	411A	225WMC-
1957	98 4F	739A	XMC-H-
	197 8E	046B	XCMC.SA
	197 9E/4	377B	200WCMC.SA
	197 9E/4	437B	WCMC.SA
	197 9E/4	377B	200XMC.SA
	197 9E/4	374B	XMC.SA
	225 1H	411A	225XMC-
	249 2T	950A	250XMC-
1958	98 4F	604B	YMC-498
	197 9E/4	437B	YMC-12 (last)
	249 2T	950A	250XMC267
	249 2T	950A	250YMC42
1959	197 9E/3	726B	YMC-292
	197 9E/3	726B	200ZMC-36
	249 2T	950A	250YMC-326
	249 2T	950A	250ZMC-99
Tandon			
1953	125 Superglide	257A	S1181
1955	197 Imp	063B	J3149-4X
1956	197 Imp/3	046B	H3416/5X
	197 Imp/4	056B	
	225 Monarch	411A	H3416/5X

183

Villiers carburettor settings

Engine	Capacity cc	Carb	Jet number	Taper needle	Set out	Groove
Midget	98	Midget	8	$5\frac{1}{2}$ Midget		
Junior	98	Junior	8J	2		
JDL	98	Junior	7J	2		
1F	99	6/0	8	$2\frac{1}{2}$	0.91	
2F	99	Junior	J8	$2\frac{1}{2}$	0.91	
4F/6F	99	{6/0 / S12}	J120 / 1	$2\frac{1}{2}$ / 2		3 / 3
9D	122	Midget	8	6		
9D	122	Lightweight	3	3		
9D	122	Lightweight	3	3 special		
10D	122	3/4	3	3	2.41	
6E	197	4/5	1	$4\frac{1}{2}$		

Engine	Capacity cc	Carb	Main jet	Pilot jet	Taper	Set out	Groove	Throttle
11D	122	S24	140	35	$3\frac{1}{2}$	1.95		2
12D	122	S19	90	35	$3\frac{1}{2}$	2.015		$2\frac{1}{2}$
29C	147	S25	130		$3\frac{1}{2}$	2.015		3
30C	147	S19	80		$3\frac{1}{2}$	2.015		$2\frac{1}{2}$
31C	148	S19	80	35	$3\frac{1}{2}$	1.97		$2\frac{1}{2}$
2L	174	S22	135	35	$3\frac{1}{2}$		4	$2\frac{1}{2}$
7E	197	S24	120	35	$3\frac{1}{2}$	1.95		3
8E	197	{S24 / S25}	120	35	$3\frac{1}{2}$	1.95		3
9E / 10E	197	S25/1	120	35	$3\frac{1}{2}$	1.95		3
1H	224	S25	120	35	$3\frac{1}{2}$	1.90		3
2H	246	S25/5	125	35	$3\frac{1}{2}$	1.90		3
31A	246	S25/1	130	35	$3\frac{1}{2}$	1.95		3 or 4
32A / 37A	246	{S25/2 / S25/5}	140	35	$3\frac{1}{2}$	1.85		4
Starmaker	247	S25/6	165	35	$3\frac{1}{2}$	2.0		3
2T	249	S22/2		35	$3\frac{1}{2}$		3	3
3T	324	S25/3	180	35	$3\frac{1}{2}$		3	
4T	249	S25/6	180	35	$3\frac{1}{2}$	1.94		3

Amal carburettor settings

Model	Year	Type	Size	Main	Pilot	Slide	Needle Pos.	Jet
AJS 250 Y4	1969–71	932	32 mm	270	20	$3\frac{1}{2}$	2	·107
AJS 380 Y5	1970–73	1034	34 mm	300	25	$2\frac{1}{2}$	3	·107
AJS 410	1973	1034	34 mm	290	25	$3\frac{1}{2}$	3	·107
Anzani 242 cc	1955–56	375	$\frac{25}{32}$	130	25	3	2	·106
Anzani comp. 242 cc	1959	376	1	300	25	$3\frac{1}{2}$	3	·109
Anzani 322 cc	1956–57	376	1	260	25	$3\frac{1}{2}$	2	·106
Corgi	1946–50	259	·425	55		3	2	·107
Corgi	1951–54	359	·425	45		3	3	·109
Cotton Cougar	1963–65	389	$1\frac{3}{16}$	480	30	3	3	·106
Cotton Cobra & Conquest	1966–67	389	$1\frac{3}{16}$	370	20	$3\frac{1}{2}$	3	·109
Cotton Telstar	1966–67	GP2	$1\frac{1}{2}$	540	25	3	1	·107
Cyc-Auto	1950–56	359	·425	55		3	3	·107
Dot 197 cc models	1950–54	261	$\frac{21}{32}$	65		3	3	·107
Dot 197 cc S7	1955–56	276	$\frac{15}{16}$	110		3	3	·107
Dot 197 cc scrambler	1960	389	$1\frac{3}{16}$	390	30	3	3	·106
Dot 250 cc scrambler	1962–65	389	$1\frac{3}{16}$	490	20	3	1	·109
Dot 250 cc scrambler	1965–67	389	$1\frac{3}{16}$	480	15	$4\frac{1}{2}$	3	·106
Dot 250 cc trials	1962	376	$1\frac{1}{16}$	400	20	$3\frac{1}{2}$	3	·106
Dot 250 cc trials	1965–67	389	$1\frac{3}{16}$	360	15	$4\frac{1}{2}$	3	·109
Dot 360 Demon	1967–68	930	30 mm	280	25	$3\frac{1}{2}$	2	·106
Excelsior 98 cc	1947–54	259	·425	55		3	3	Std
Excelsior 98 cc Minor	1947–52	259	·531	70		5	2	·107
Excelsior 98 cc Spryt	1955–56	359	·425	55		3	3	·107
Excelsior 125 cc Minor	1949–52	261	$\frac{5}{8}$	75		3	3	·107
Excelsior 150 cc Minor	1951–52	274	$\frac{25}{32}$	80		5	2	Std
Excelsior 150 cc Courier	1953	275	$\frac{7}{8}$	90		4	3	Std
Excelsior 150 cc Courier	1955–56	375	$\frac{7}{8}$	140	30	$3\frac{1}{2}$	2	·105
Excelsior 150 cc Courier	1959–60	376	1	180	25	4	3	·106
Excelsior 250 cc Twin	1950–55	274	$\frac{25}{32}$	80		5	2	Std
Excelsior 250 cc Twin	1955–60	375	$\frac{25}{32}$	100	30	3	2	·105
Excelsior STT6	1958	375	$\frac{7}{8}$	150	30	4	2	·105
Excelsior 328 cc Twin	1958	376	1	230	25	4	2	·106
Francis-Barnett 125 cc	1951	205	$\frac{7}{8}$	110		4	3	·107

Part Three Appendix

Model	Year	Type	Size	Main	Pilot	Slide	Needle Pos.	Jet
Francis-Barnett 125 cc	1953–54	276	$\frac{15}{16}$	110		3	3	·107
Francis-Barnett 150 cc 15T	1958–66	375	$\frac{13}{16}$	110	25	$3\frac{1}{2}$	2	·105
Francis-Barnett 175 cc 79	1956–59	370	$\frac{15}{16}$	240	25	4	3	·106
Francis-Barnett 197 Falcon	1950–58	276	$\frac{15}{16}$	110		3	3	·107
Francis-Barnett 197 Falcon	1953–57	276	1	130		3	3	·107
Francis-Barnett 197 Falcon	1960–66	376	1	180	30	$3\frac{1}{2}$	2	·106
Francis-Barnett 250 Cruiser	1957	389	$1\frac{1}{8}$	420	30	3	2	·105
Francis-Barnett 250 Cruiser	1958–63	389	$1\frac{1}{8}$	320	30	$3\frac{1}{2}$	2	·105
Francis-Barnett 250 trials	1958–61	376	1	190	25	$3\frac{1}{2}$	3	·1055
Francis-Barnett 250 scrambler	1958–59	389	$1\frac{1}{8}$	290	30	4	2	·105
Francis-Barnett 250 scrambler	1966	389	$1\frac{3}{16}$	370	20	$3\frac{1}{2}$	3	·109
Greeves 197 cc M-X	1958–60	376	$1\frac{1}{16}$	280	25	3	3	·106
Greeves 197 cc M-X	1958–60	389	$1\frac{3}{16}$	440	30	3	3	·106
Greeves 242 cc 25D	1956–58	375	$\frac{25}{32}$	130	25	3	2	·106
Greeves 322 cc 32D	1955–57	376	1	260	25	$3\frac{1}{2}$	2	·106
Greeves 197 cc M-X	1963–65	389	$1\frac{3}{16}$	460	25	3	3	·109
Greeves 250 cc M-X	1964–66	389	$1\frac{3}{16}$	350	25	$3\frac{1}{2}$	2	·106
Greeves 250 cc M-X/trials	1963–65	389	$1\frac{3}{16}$	380	25	$3\frac{1}{2}$	4	·109
Greeves 250 cc M-X	1963	389	$1\frac{3}{16}$	370	25	$3\frac{1}{2}$	3	·109
Greeves 250 cc M-X	1963	pair 389	$1\frac{1}{8}$	290/210	30	3	5/3	·106
Greeves 250 cc trials	1966	376	$1\frac{1}{16}$	190	25	3	3	·105
Greeves Silverstone	1963–68	GP2	$1\frac{3}{8}$	390	25	3	2	·109
Greeves 250 cc 24MX	1967–68	930	30 mm	280	20	3	3	·106
Greeves 250 cc 24TH	1967–68	626	26 mm	140	25	$2\frac{1}{2}$	2	·106
Greeves 362 cc 36MX	1967–68	932	32 mm	300	30	3	2	·106
Greeves 250 cc	1969–74	932	32 mm	310	25	3	2	·106
Greeves 380 cc	1972–74	1034	34 mm	320	30	3	2	·107
James 150 cc	1958–66	375	$\frac{13}{16}$	110	25	$3\frac{1}{2}$	2	·105
James 175 cc	1956–59	370	$\frac{15}{16}$	240	25	4	3	·106
James 197 cc trials	1956	276	$\frac{15}{16}$	140		3	3	·107
James 197 cc comp.	1957	376	$\frac{15}{16}$	180	20	$3\frac{1}{2}$	2	·105
James 199 cc	1960–66	376	1	180	30	$3\frac{1}{2}$	2	·106
James Commodore	1957–61	389	$1\frac{1}{8}$	320	30	$3\frac{1}{2}$	2	·105
James Commodore	1957–58	389	$1\frac{1}{8}$	420	30	3	2	·105
James 250 cc trials	1958–59	376	1	190	25	$3\frac{1}{2}$	3	·1055
James 250 cc trials	1966	389	$1\frac{3}{16}$	370	20	$3\frac{1}{2}$	3	·109

186

Villiers Singles & Twins

Model	Year	Type	Size	Main	Pilot	Slide	Needle Pos.	Jet
Norman 150	1955–57	276	$\frac{15}{16}$	110		3	3	·107
Norman 197 cc	1952–57	276	$\frac{15}{16}$	110		3	3	·107
Norman 242 cc twin	1955–58	375	$\frac{25}{32}$	130	25	3	2	·106
RCA 350 cc	1959	376	$1\frac{1}{16}$	180	20	$3\frac{1}{2}$	3	·106
RCA 350 cc	1960	376	$1\frac{1}{16}$	170 or 220	20	$4\frac{1}{2}$	2	·106
Tandon 197 cc	1955–58	276	1	130		3	3	·107
Tandon 242 cc twin	1955–56	375	$\frac{25}{32}$	130	25	3	2	·106
Tandon 322 cc twin	1955–56	376	1	260	25	$3\frac{1}{2}$	2	·106
Villiers 9E	1962–66	376	$1\frac{1}{16}$	290	30	3	4	·106
Villiers 250	1959–60	389	$1\frac{3}{16}$	480 or 560	30	3	4 or 5	·106
Villiers 33A, 34A, 36A	1962–64	389	$1\frac{3}{16}$	370	25	$3\frac{1}{2}$	3	·109
Villiers Starmaker	1963–65	389	$1\frac{1}{8}$	430		3	3	·106
Villiers Starmaker	1963–66	389	$1\frac{1}{8}$	460	30	3	5	·106
Villiers Starmaker racer	1964	GP2	$1\frac{1}{2}$	420	30	4	3	·109
Villiers Starmaker racer	1965–66	GP2	$1\frac{1}{2}$	540	25	3	5	·107
Villiers Starmaker M-X	1964–66	389	$1\frac{3}{16}$	370	20	$3\frac{1}{2}$	3	·109
Villiers Starmaker M-X	1967–68	932	32 mm	260	25	3	2	·106
Villiers Starmaker	1967–68	932	32 mm	300	25	3	2	·106

"Oh, just cuff him on the ear!"

Corgi cartoon of 1948